Oilopoly

D0494055

"A gripping, highly topical, and important book – and what a rattling good read!"

Richard Sakwa, *Associate Fellow of the Russia and Eurasia Programme at the Royal Institute of International Affairs*

"A superb, readable description of Vladimir Putin's role in the emergence of Russia as a successful and potentially threatening petrostate."

James R. Millar, *Professor of Economics and International Affairs, George Washington University*

"This may be Goldman's best book, and that's saying a lot. Focusing on Putin's Russia with a scholar's commitment to deep and meaningful research and a reporter's eye for detail and colour, Goldman has explained why and how Russia has again emerged as a global power. The answer is oil. At inflated prices, it leads directly to inflated national aspirations and further down the road to dangers of a totally unpredictable nature. Read and learn."

Marvin Kalb, *former Moscow bureau chief for CBS News*

"Written in an engaging, lively, and accessible style, *Oilopoly* vividly outlines the importance of the Russian oil and gas industry. It masters the complex development of the history of Russian energy policy and compellingly relates to Putin's political ambitions."

D   ane, *editor and contributor to* The Political Economy of Russian Oil

"Few developments are likely to reshape the contours of international politics the next decade more than Russia's ascent to energy superpower. And one can tell the story of that ascent and the challenges it presents with greater knowledge or flair for detail than Marshall Goldman."

Mark R. Beissinger, *Professor of Politics, Princeton University*

"A timely and sobering new study of Moscow's petroleum industry. Putin is at the centre of Goldman's readable study of the resurgence of Russian power based on petro-dollars. But the author combines sound history with economic analysis to come to the important conclusion that the new assertiveness of the Kremlin is here to stay."

Norman M. Naimark, *Robert and Florence McDonnell Professor, Stanford University*

Marshall Goldman is Professor of Economics Emeritus at Wellesley College and Senior Scholar at the Davis Center for Russian Studies at Harvard University. An internationally recognized authority on Russian history, politics, and economics, he has met with Mikhail Gorbachev and Vladimir Putin, and has advised former President George H.W. Bush and President George W. Bush on Russia.

Goldman's many other publications include *Lost Opportunity: What has Made Economic Reform in Russia so Difficult?* and *The Piratization of Russia: Russian Reform Goes Awry*. He has also written for such publications as *Foreign Affairs*, *The New York Times*, *The Washington Post*, and *The New Yorker*, and is a regular contributor to the Russian newspapers *Moscow News* and *The Moscow Times*.

Marshall Goldman and his wife, Merle, a Professor Emerita of Chinese History at Boston University, are the parents of four children and currently reside in Wellesley, Massachusetts.

OILOPOLY
Putin, Power and the Rise of the New Russia

Marshall Goldman

ONEWORLD
OXFORD

A Oneworld Book

First published in Great Britain by Oneworld Publications 2008

Copyright © Marshall Goldman 2008

The right of Marshall Goldman to be identified as
the author of this work has been asserted by him
in accordance with the Copyright, Designs and
Patents Act 1988

ISBN 978 1 85168 621 6 (Hbk)
ISBN 978 1 85168 646 9 (Pbk)

Cover design by designedbydavid.co.uk
Printed and bound in Great Britain by Bell & Bain, Glasgow

Oneworld Publications
185 Banbury Road
Oxford OX2 7AR
England
www.oneworld-publications.com

Learn more about Oneworld. Join our mailing list to
find out about our latest titles and special offers at:

www.oneworld-publications.com

To Merle; not the First Dedication and I Hope not the Last

Contents

Figures and Tables

FIGURES

TABLES

Preface

More than in my past writing efforts, I owe thanks to a set of enthusiastic helpers. They provided invaluable help in preparing my manuscript. Two of them have the ability to read my handwriting, something I am not always able to do myself. Doing my best to ignore the advances of the modern computerized world, I prefer to write out the text in longhand on legal-size yellow pads. Robert Price was able to transcribe those writings for me onto a computer, so I was devastated when he went to work at a higher calling. To my relief Sue Sypko took over and proved to be as able, and, equally important, she hasn't frowned when I bring her yet another set of nearly incomprehensible scribbles. In fact, I have taken to awarding her Stakhanovite prizes for her efforts. The third member is Coco Downey, who offered herself as research assistant and eagerly agreed to chase after obscure facts and display them in a way that aids the understanding of how things work in Russia. I have come to call her "the wizard." After reading her charts and diagrams in the chapters that follow, I suspect the readers, even those in Russia, will agree that they can now understand the previously incomprehensible. The fourth and most unlikely member of this quartet is Thomas Luly, a most amazing high school junior. Out of the blue he wrote an e-mail asking if I needed any assistance. To humor him, I sent him an early draft of the manuscript and to my amazement, he not only read the whole thing and made extensive notes, but he found more inconsistencies in the text than I am embarrassed to admit should have been there. He also asked some probing questions that should help both me and I hope future readers deal with issues that are all too often skirted.

I am indebted to all four of these collaborators, Robert, Sue, Coco, and Thomas.

John D. Grace also deserves a special note of thanks. He read the manuscript with admirable care and made some especially valuable suggestions, almost all of which I have incorporated. Of course I am ultimately responsible for whatever mistakes remain, but he and the gang of four spared me from many others.

Then there are others to whom I must also express my thanks. The RIA Novosti Press Agency provided me, as part of the Valdai Hills Discussion group, with the opportunity to meet with President Vladimir Putin on four occasions extending over an extraordinary twelve hours. They also took us out to the Priobskaia oil fields and Yuganskneftegaz and arranged a meeting at Gazprom headquarters. It was as if I had died and gone to heaven. They let me come along although invariably I asked the least respectful questions.

I must also thank Kathryn Davis, whose chair I held at Wellesley College, for her interest (even at her 100th birthday party) and for her financial support. She has helped to reassure me that there was always someone out there who was as interested in Russia and its sometimes troubling ways as I was. Her son Shelby and daughter Diana share much of that same enthusiasm.

Most of all, of course, I must also thank my wife, Merle. She has to put up with a lot, enough in fact to tighten anyone's digestive system. While I often ask myself if I can withstand another of her brutal, yes, brutal, editing jobs, in the end I have to concede, but not directly, that I and the manuscript are better off for it. But after fifty-five years, it is a testimony to the strength of our marriage that we have survived a joint husband and wife writing and editing effort. There aren't many couples I know of who can say the same thing, but Merle is special, and our children and I can never acknowledge how much we owe her.

Introduction

Russia—Once Again an Energy Superpower

THE AUTHOR AS JAMES BOND

At first I was puzzled. Where were they taking us? For such a big, sleek, glass Moscow high rise, Gazprom's elevator in its headquarters building was tiny (five people could barely squeeze in) and its hall corridors narrow. This was, after all, the world's largest producer of natural gas, not to mention Russia's largest company. Following a short walk we were ushered into a darkened, silent room where nothing seemed to be happening. Strange.

It was only when all the members of our group had made their way up on the elevators that the room suddenly came alive. Then for a time I felt as if I had wandered into the NASA Space Center, or was it a James Bond movie set? All that was missing was that out of body voice intoning, "Welcome, Mr. Goldman. We were expecting you."

In front of me, covering the whole 100-foot wall of the room, was a map with a spiderweb-like maze of natural gas pipelines reaching from East Siberia west to the Atlantic Ocean and from the Arctic ocean south to the Caspian and Black Seas. Manipulating this display were Gazprom dispatchers, three men controlling the flow of Gazprom's gas to East and West European consumers of this Russian natural gas monopoly. No wonder there was tight security. There was also a sense of self-assurance. As measured by the value of its corporate stock, by summer 2006, Gazprom, this state-dominated joint stock corporation (until 1992 it was actually the Soviet Ministry of the Gas Industry), had become the world's third-largest corporation. Only private shareholder-owned Exxon-Mobil and General Electric were larger.

With a flick of a switch, those dispatchers sitting in this Moscow room could freeze—and indeed have frozen—entire countries. At the very least, they could send their citizens off in a panic in search of sweaters, scarves, and blankets. What an empowering feeling! Should they choose to, those Gazprom functionaries could not only cut off natural gas from the furnaces and stoves of 40 percent of Germany's homes but also the natural gas that many German factories need for manufacturing a range of products from ammonia fertilizer to plastics. While Germany purchases more natural gas from Russia than any other country in Europe, all of Western Europe is now also hooked up directly or indirectly to the Gazprom pipeline. In the extreme case, the Baltic states and Finland import 100 percent of their natural gas from Russia.

Here then in front of me was the natural gas distribution brain center for virtually the whole European continent. I could not think of anything comparable in Europe where such an essential commodity can be controlled by one country, and more than that, one company. In this very room the dispatchers factor in weather forecasts, special production needs, holidays, and, while they are reluctant to acknowledge it (in fact they deny it), a customer's political correctness. Despite the fact that the Russian government owns more than 50 percent of Gazprom's shares and President Vladimir Putin takes a very personal

FIGURE I Dispatching Center at Gazprom Headquarters, Moscow. Gazprom is Russia's largest company and the biggest extractor of natural gas in the world. Copyright © Mauro Galligani/Contrasto-Redux.

and intense interest in Gazprom's operations, Gazprom officials insist that politics never, ever affect their calculations.

"Gazprom is a reliable energy partner" goes the mantra: it adheres to its contracts, guarantees delivery, and assures "energy security." As Alexander Medvedev, the deputy chairman of Gazprom, told us that same morning, "What is good for a strong Gazprom is good for the world." Reminiscent of Charles E. Wilson, the CEO of General Motors in the 1950s who boasted that "What was good for our country was good for General Motors and vice versa," Medvedev's pairing of Gazprom and the world is understandable but is as much off the mark as was Charlie Wilson's earlier formulation.

Russia has not hesitated in the past to cut off the flow of both petroleum and gas to strengthen its side of a political dispute, a practice it inherited from its forebears in the Soviet Union's Ministry of the Gas Industry and Ministry of the Petroleum Industry. Europeans are realizing how dependent on Russia they have become as each year they rely more and more on Russian natural gas imports. Gazprom and, by extension, the Russian government are already beginning to enjoy a power over their European neighbors far beyond the dreams of the former Romanov czars or the Communist Party general secretaries. President Vladimir Putin, with his control of Gazprom as well as another state-owned petroleum company, Rosneft, had become a real-life Dr. No—an archetypal James Bond villain, complete with a yacht and retinue. As President Putin at the time noted in a three-hour meeting following our Gazprom visit, Gazprom and Rosneft are very real and each year are accumulating more and more wealth and international influence, which they are using to advance the interests of the Russian state.

But it is not only Europe that finds itself each day becoming more and more dependent on energy exports from Russia. Although the United States is separated from Russia by oceans, it also is beginning to import and consume more and more Russian energy. As in Europe, the United States is trying to reduce its overreliance on energy imports from the Middle East. As part of this diversification, in 2005 the United States imported close to $8 billion worth of Russian petroleum. In 2006, that jumped by 25 percent to $10 billion. True, that represented only 3 percent of U.S. petroleum imports—small, but an increase from the 2.2 percent of 2004 and a hint that they are likely to increase imports in the future.[1] More than that, in 2000, LUKoil, one of Russia's largest private oil companies, purchased nearly 3,000 filling stations in the United States from Getty Oil and Mobil and is now busily converting them into LUKoil outlets. It also should be noted that in 2006, Russia became the world's largest producer of petroleum, producing more than Saudi Arabia. This is not the first time Russia has produced more petroleum than anyone else. It also reigned as the world's largest producer in the late 1970s and 1980s. Even this was not unprecedented. As Table Intro.1 indicates, Czarist Russia from 1898 to 1901 also produced more oil than the United States, until then the leader.

Equally unusual, even though there are no natural gas pipelines connecting the United States with Russia, Gazprom is also beginning to export LNG (liquified natural gas) to the United States. For the time being, because Gazprom as yet lacks the technology to produce LNG on its own, it is a swap arrangement. These shipments under the Gazprom label actually originate in Algeria (in exchange, Gazprom pipes gas to some of Algeria's customers in Europe), but by 2010, Gazprom anticipates (unrealistically) that it will supply as much as

TABLE INTRO.1   Russian and American Petroleum Production and Exports (mill. metric tons)

	Russian Production	Russian Export	US Production
1860	0.004		0.068
1861	0.004		0.273
1862	0.004		0.409
1863	0.006		0.356
1864	0.009		0.288
1865	0.009	0.002	0.341
1866	0.013	0.002	0.491
1867	0.017	0.003	0.466
1868	0.029	0.002	0.497
1869	0.042	0.001	0.575
1870	0.033	0.002	0.715
1871	0.026	0.001	0.710
1872	0.027	0.002	0.858
1873	0.068	0.001	1.350
1874	0.106	0.002	1.490
1875	0.153	0.002	1.660
1876	0.213	0.002	1.242
1877	0.276	0.001	1.820
1878	0.358	0.001	2.100
1879	0.431	0.005	2.716
1880	0.382	0.003	3.575
1881	0.701	0.018	3.76
1882	0.870	0.019	4.128

	Russian Production	Russian Export	US Production
1883	1.039	0.059	3.189
1884	1.533	0.113	3.294
1885	1.966	0.178	2.973
1886	1.936	0.247	3.817
1887	2.405	0.311	3.846
1888	3.074	0.573	3.755
1889	3.349	0.734	4.782
1890	3.864	0.788	6.232
1891	4.610	0.889	7.384
1892	4.775	0.937	6.870
1983	5.620	0.985	6.587
1894	5.040	0.880	6.710
1895	6.935	1.059	7.193
1896	7.115	1.058	8.290
1897	7.566	1.046	8.225
1898	8.635	1.115	7.530
1899	9.264	1.392	7.762
1900	10.684	1.442	8.652
1901	11.987	1.559	9.468
1902	11.621	1.535	12.072
1903	11.099	1.784	13.662
1904	11.665	1.837	15.923
1905	8.310	0.945	18.322
1906	8.885	0.661	17.203
1907	9.760	0.733	22.589
1908	10.388	0.797	24.280
1909	11.248	0.796	24.911
1910	11.283	0.859	28.500
1911	10.547	0.855	29.981
1912	10.408	0.839	30.319
1913	10.281	0.948	36.144
1914	10.013	0.529	38.230
1915	10.138	0.078	40.904
1916	10.8		41
1917	8.8		45.7
1918	4.1	2	49
1919	4.4		52
1920	3.9		60

(continued)

	Russian Production	Russian Export	US Production
1921	3.8	7	64
...			
1929	13.7	3.9	137
1930	18.5	4.7	122
1931	22.4	5.2	116
1932	21.4	6.1	103
1933	21.5	4.9	123
1934	24.2	4.3	124
1935	25.2	3.4	136
1936	27.4	2.7	150
1937	28.5	1.9	174
1938	30.2	1.4	166
1939	30.3	0.4	172
1940	31.1	0.9	185
...			
1946	21.7	0.5	237
1947	26	0.8	253
1948	29.2	0.7	275
1949	33.4	0.9	251
1950	37.9	1.1	269
1951	42.3	2.5	306
1952	47.3	3.1	312
1953	52.8	4.2	321
1954	59.3	6.5	316
1955	70.8	8	339
1956	83.8	10.1	357
1957	98.3	13.7	356
1958	113.2	18.1	334
1959	129.6	25.4	351
1960	147.9	33.2	351
1961	166.1	41.2	357
1962	186.2	45.4	365
1963	206.1	51.3	375
1964	223.6	56.6	380

Data from Vneshni Torgovli Rossii, Department tamozhenny sbor' (for various years) and U.S. Bureau of the Census, Historical Statistics of the United States, Colonial Times to 1970.

10 percent of the natural gas the United States needs as LNG directly from its own fields.[2] Given that the United States has fairly large natural gas reserves of its own and supplements domestic production with imports by pipeline from Canada, it is unlikely that the United States will ever become as beholden to Russia for its energy as Germany or Austria have become. Yet Russia's emergence as an energy superpower will have a long-term impact on U.S. and world diplomacy if for no other reason than that our European allies will begin to think twice before saying "no" to Russia.

EUROPE BECOMES VULNERABLE

I had a chance to discuss this new strategic relationship with President George W. Bush at a June 2006 meeting in the Oval Office. The meeting was called to brief him before the G-8 meeting in St. Petersburg, which was to take place a few weeks later. President Putin as chairman of the 2006 meetings had repeatedly insisted that energy security and Russia's role as a reliable supplier should be the general theme. "The Russian Federation has always abided by all of its obligations, fully and completely, and it will continue to do so."[3] Considering that only seven months earlier during a cold January 2006, Gazprom had curbed its flow of gas to Ukraine, which in turn reduced the flow to the rest of Western Europe, this was a rather dubious concept. Ostensibly, Gazprom justified the drop in gas exports to Ukraine, as well as to Georgia and Moldova, explaining that it did this because all three refused to pay the European market price. But since other Russian customers, especially Belarus and Armenia at the time, were also paying below European market prices, it was widely agreed that Russia was using its gas more as a political than an economic weapon.

As Russia's customers have awakened to how vulnerable they have become to future cuts in their energy supplies, there are signs that this overdependence on Russian gas is already forcing at least some in Europe to have second thoughts about standing up to Russia. Nor are the European consumers the only ones who find themselves very much at the mercy of Gazprom. So far Gazprom also determines the fate of three other large exporters of natural gas. To their dismay, if they want to sell natural gas to Europe, Central Asian gas producers such as Turkmenistan, Kazakhstan, and Uzbekistan have no alternative but to ship it through the Gazprom pipeline. This is a legacy of the Soviet era when it was only logical to consolidate shipments of gas produced

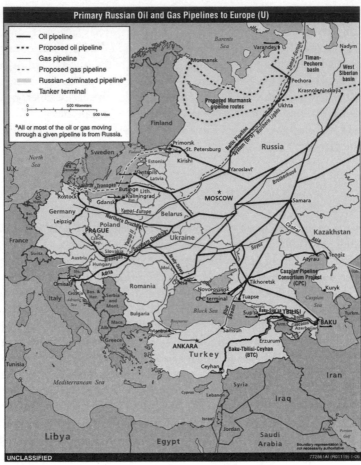

FIGURE 2   Primary Russian Oil and Gas Pipelines to Europe (U).
Source: EIA.

within the republics of the Soviet Union through one unified system.
After all, what did it matter if the gas to be exported came from
Uzbekistan and transited through Russia? They were both parts of the
Soviet Union. But when the Soviet Union disintegrated in 1991,
Gazprom assumed ownership of the bulk of that pipeline, and the
newly independent countries in Central Asia, which were previously
republics of the USSR, had no other outlet of their own to the West.
As a result, this post-1991 monopoly control of the natural gas pipeline
allows Gazprom to hold down the price it pays to the Central Asian
producers for their gas. In 2006, for example, Gazprom paid less than

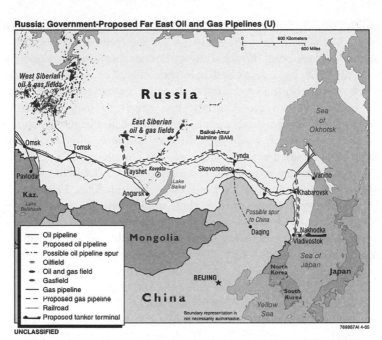

Russia: Government-Proposed Far East Oil and Gas Pipelines (U)

Legend:
— Oil pipeline
- - Proposed oil pipeline
-·- Possible oil pipeline spur
☰ Oilfield
● Oil and gas field
● Gasfield
— Gas pipeline
- - Proposed gas pipeline
+++ Railroad
---- Proposed tanker terminal

UNCLASSIFIED

769957AI 4-05

FIGURE 3 Russia: Government Proposed Far-East Oil and Gas Pipelines (U). Source: EIA

$50 per 1,000 cubic meters while selling this same gas to the Europeans at prices averaging $230 per 1,000 cubic meters.

Working with energy and government officials in Central Asia, some European countries and the United States have sought to break that monopoly by lining up support for a bypass natural gas pipeline that would be built under the Caspian Sea and link Central Asia to Azerbaijan. There it would parallel an oil pipeline that goes then to Georgia and Turkey. To be called NABUCCO, this pipeline would then run through Bulgaria, Romania, Hungary, and on eventually to Western Europe.

To prevent any such dilution of their pipeline monopoly, Putin and the Russians have homed in on Hungary and worked to convince it to back the Russian alternatives. To draw them away from NABUCCO, the Russians proposed that if they supported the Russian alternative version, Hungary would become the Central European hub for redistribution of natural gas to the rest of Europe. By contrast, in the NABUCCO version that excludes Russian natural gas, Austria would be the hub. In a further counterproposal, Russia promised that if Hungary held back and signed up with the Russians, not with NABUCCO, it would be well provided for with assured natural gas deliveries for the foreseeable future. No doubt this is tempting. While Hungary has not disavowed participation in NABUCCO and may yet back it, by the spring of 2007, the Hungarian prime minister began to speak favorably of the Russian pipeline version.[4] By mid-2007, even the Austrians had begun to back away from NABUCCO. And by January 2008, Bulgaria had also begun to support South Stream, which would be a Russian-sponsored alternative. The Russians understood well that there was still not enough of a market to justify the hundreds of millions, if not billions, of dollars that would be needed to build two, much less three, competing pipelines. Thus, should Hungary opt for a Russian variant, there would be too few customers for NABUCCO, which would make it unprofitable. This would make it impossible to attract the necessary investment, which would spell all but certain death for NABUCCO. This Russian maneuver is a good example of the skillful chess game Putin and his subordinates are now playing, using their natural gas and oil as their rook and queen.

While President Putin disavows any notion that today's Russia has become an energy superpower, in reality it has. Yet these oil and gas resources are not newly discovered.[5] However, the challenge has always been to harness those resources and use them effectively. After all, as Russia is the largest country in the world geographically (eleven time zones), it was inevitable that under some of those Russian hectares there would be large deposits of crude oil and natural gas. But Russia has often had trouble locating those deposits, bringing them to the surface, and then transporting them to domestic and foreign consumers. Given the northern latitude and the offshore location where so many of Russia's energy deposits are located, as well as the distance from foreign consumers, this has not been an easy task. Russian winters are long and very cold (ask Napoleon and Hitler who had tried to conquer that country), and the summers are short, often very warm, and because of the permafrost blanketing much of the north, almost always swampy and full of mosquitoes. As for transportation, there are few or no easily accessible warm water or deep seaports. Nor does it help that the rivers in Siberia almost all run north to the Arctic, not east or west to the most populated areas where both the domestic and foreign consumers of that oil and gas reside.

It is no wonder then that the development of Russia's energy resources has been belated, challenging, and intermittent. To complicate the effort even more, Russian drilling technology has often lagged behind that of the rest of the world, particularly the type used in offshore deepwater drilling. The gap in technology was especially harmful during the Soviet period when out of fear of ideological contamination, the Soviet Union prohibited many of its enterprises and technicians from having access to the West. At the same time, led by the United States, many Western governments withheld their advanced technology from Russia.

HISTORY REPEATS ITSELF

Given this background, the struggle for access and control of Russian energy resources provides an often overlooked and therefore neglected perspective as to why Russia, be it in the current, Czarist, or Soviet eras, developed as it did. Although not widely known, Russia has led the world in the production of petroleum several times in its history, despite so many difficulties. As we just saw, from 1898 to 1901 Russia outproduced the United States, until then the world's largest producer. The United States resumed first place thereafter and remained the world leader for seven decades until 1975. (See Table 2.1.) But while U.S. production generally began to decline, petroleum output in the USSR began to increase at annual rates of 5–6 percent. By 1975, the USSR again outproduced the United States and thus again became the world's largest producer. The Russian Republic alone when it was a part of the USSR produced more than the United States in 1980. Even after the collapse of the USSR, if only for a brief time, Russia remained the world's largest producer. However, in 1992, one year after the breakup of the USSR, Saudi Arabia's output exceeded Russia's. The disintegration of the USSR and the confusion and economic and political freefall that followed precipitated a sharp drop in Russian output. By 1996 production was 45 percent below what it had been in 1990. In 1999, crude oil output began to increase again, but Saudi Arabia continued to outproduce both Russia and the United States until 2006. Then once again in 2007 Russia regained its place as the world's largest producer. (See Table 2.1.)

As Russia's petroleum production statistics suggest, petroleum has played an important if not crucial role in Russia's economic and political life. But just as in other resource-rich countries, this role has not always been a positive one. Unless, as in the United States or Norway, there is an already established integrated market industrial infrastructure in place, there is a danger—as most if not all of the OPEC members have discovered—that relying on that oil and natural gas as the main export earners can corrupt the country. The availability of large deposits of petroleum and natural gas tends to bring with it an overreliance on those resources at the expense of more labor-intensive manufacturing and the development of technology and human capital.

RUSSIA—A VICTIM OF THE DUTCH DISEASE

No doubt the size of a country's resource endowment does make a difference in the way that country develops. Economists often debate whether countries such as Japan, South Korea, Taiwan, and Switzerland would have developed industrially and technologically as they did if they had been more richly endowed. Conversely, if in 1917 the Germans had sent Lenin in that sealed train to Tokyo instead of Petrograd, it is most likely that the communist country that he would have created there would have ended up with a very different incentive system than the one adopted by what would become the Soviet Union. Because resources were so abundant in Russia, Soviet leaders set very low prices on their metals and fuels. Given the scarcity of such resources in Japan, the odds are that in a hypothetical Soviet Japan, raw material prices would have been much higher than they were in a Soviet Russia, reflecting that scarcity. And if Japan, South Korea, Taiwan, and Switzerland today were suddenly to discover new and abundant deposits of oil and gas, it is probable that, just like Russia, they too would price their raw materials cheaply and thus would also be afflicted with the Dutch disease, so called because once the Dutch found natural gas off their North Sea Coast, the relative prosperity it brought came at the expense of the country's manufacturing sector. The export of that gas created a heavy demand for the Dutch guilder that the foreign buyers needed to pay for their purchases. This pushed up the value of the guilder. With a stronger currency, the citizens of the Netherlands found that imported goods were now cheaper than they were before as well as cheaper compared to goods manufactured within the Netherlands itself. The strong guilder also made Dutch exports more expensive for foreigners. Inevitably this had an adverse impact on domestic manufacturing and resulted in a loss of manufacturing jobs in Dutch factories.

The increase in the value of the ruble relative to other currencies—precipitated by the increase in both the price of a barrel of oil and the sharp increase in production and the export of Russian petroleum after 1999—also gave rise to what can be called the Russian Disease.[6] Not only does a booming export market for energy resources have an adverse impact on domestic manufacturers but the appearance of a large and expanding petroleum sector inevitably triggers a ferocious struggle to win control of those oil-producing fields, at least in countries where the state allows the private ownership of energy-producing companies. Related to this struggle for control, whenever petroleum and gas industries begin to dominate a country's economy, democratic institutions often seem to weaken if not collapse. Venezuela is one of the more recent examples.

Is a large petroleum and natural gas endowment a blessing or a curse? There is no all encompassing answer to such a question. In Norway, where the discovery of natural gas came many years after the country had already been industrialized, the disruption has been relatively minor. That is because the Norwegians understood that the sudden influx of energy export revenue can have a very negative effect on both the economy and the moral makeup of the country. For that reason the Norwegians have made a determined effort to shelter the rest of the economy from this energy windfall. They have set aside export revenues in a special fund to hold down inflation and prevent their currency from gaining too much value. So far they seem to have succeeded. As a result Norway's oil and gas deposits have not become a curse. By contrast, it is hard to see how the average citizen in countries such as Libya, Iran, Nigeria, or even some in Saudi Arabia has benefited from his or her country's energy riches.

How have its energy riches affected Russia? Given the predominance of energy in the makeup of both its GDP (about 30 percent) and its exports (almost 65 percent of the 2006 total), it might initially appear that, like Saudi Arabia, energy in Russia has had a similarly adverse impact on the effort to build up Russian industry. But this overlooks the fact that Russian industry has never been a competitive force in world markets comparable to industry in most of Europe, Japan, or the United States. Industry in the Czarist era before the Bolshevik Revolution was only just beginning to respond to domestic needs. In the Soviet years, the development of a domestic industry was a major goal of the central plan era and the Soviets did indeed create new industries. Yet after the disintegration of the USSR in 1991, it quickly became apparent that the domestic Russian industry created during the Soviet era was essentially of the hothouse variety, designed primarily to build up the country's military-industrial complex. Such industries usually have a hard time when forced to sell in international markets, and those Russian manufacturers rarely were able to succeed on a purely competitive basis. Russian energy resources were used more as a lifeboat to support a non-market-oriented economy and the country's industrial dinosaurs rather than as a stimulant to growth and the development of a world-class competitive manufacturing complex.

RUSSIA SUFFERS AND RECOVERS

Yet in contrast to its sometimes negative domestic economic, political, and social impact during both the Soviet and post-Soviet eras, Russian energy has played a major role in enhancing Russia's international political standing. In many cases, it is almost as important as the development of the Soviet Union's military capabilities. Energy exports opened up doors to Soviet influence in much of the third world prior to 1991, Cuba being the best example. But in a repeat of earlier burst bubbles, with the onset of the energy glut in the late 1980s and throughout the 1990s, Russian energy became irrelevant in the world's energy balance. With production and exports down by almost 50 percent and crude oil prices hovering at a low of $10–$12 a barrel, Russia had trouble paying its bills and as a consequence suffered a massive financial collapse. On August 17, 1998, the government defaulted on its debt and most of the country's private banks closed their doors and locked their vaults. As a result, not only the banks but the country as a whole teetered on the edge of bankruptcy. The ruble lost most of its value. But in a remarkably quick turnaround, in 1999 the global demand for energy began to outpace the producers' ability to respond to that demand. As before in its history, as world energy markets quickly absorbed their spare capacity, Russia's petroleum and gas suddenly took on a new importance, economically and especially politically. Fueled by petroleum prices for Brent oil that at one point in 2005 exceeded $70 a barrel, Russian companies responded by sharply increasing production. Forty percent of the world's increased petroleum consumption from 2000 to 2004 came from Russia. As a result Russia found itself inundated not only with dollars and euros but with political leverage that in many respects exceeded anything enjoyed in either the Czarist or Soviet eras.

True, Russia may no longer be a military world superpower, but there is little doubt that despite President Putin's insistence that it is not one, Russia today is again a superpower. Only now it is an *energy* superpower.

Nor are Putin and those around him leaving this to chance. At first glance it may seem that much of this is just a matter of luck. But as we shall see, a more careful examination shows that this use of the country's natural resources and the way they are exploited by what Putin has come to call "national champions" is all part of a carefully thought out grand strategy. Part of that strategy calls for the reimposition not only of state control but of state ownership (renationalization) of at least 50 percent plus one share of the stock of many of the petroleum, metal, and manufacturing companies that were privatized in the mid-1990s. Led by Rosneft where the state has always held majority ownership, companies like Yukos and Sibneft have been effectively renationalized. (How far-reaching this has been we will see in greater detail in Chapter 5, Table 5.4.) That explains why the share of crude oil production produced by the state-dominated companies in the year 2000, the year Putin took over as president, had fallen to as low as 10 percent. However, by 2007, just before he gave up the presidency, state-dominated companies' share of crude oil production had risen again to close to 50 percent.

With its natural gas and oil pipelines that tie Europe to Russia like an umbilical cord, Russia has unchecked powers and influence that in a real sense exceed the military power and influence it had in the Cold War. No matter how many nuclear weapons it may have had, the USSR was prevented from using them by the knowledge that the United States had a comparable number and would counter the USSR's use of them and vice versa. This was referred to as Mutually Assured Destruction (MAD), which meant no one country would dare attack the other. Now, however, if Russia decides to reduce or suspend the flow of gas through its pipeline to Ukraine and/or to Europe, there is virtually nothing to restrain it from doing so. There is no comparable Mutual Assured Restraint or MAR. It is also noteworthy that this gives Russia more economic clout with Europe than Saudi Arabia. Because the Saudis export relatively little natural gas, there are no consuming countries dependent on a Saudi pipeline for this commodity. This is an important strategic difference.

In the pages that follow, we will see how the ups and downs of the Russian energy sector provide a unique insight into what is taking place in the country as a whole. Our account should also help us understand some Russian paradoxes. Given that in the late 1970s and early 1980s, Russia (then part of the USSR) was the world's largest producer of petroleum, why in 1991, with all that mineral wealth, did the Soviet Union collapse? What role, if any, did the CIA play in that collapse? Why wasn't Russia an energy superpower then and why is it now? What are the implications of all this for Europe and the United States? How much of this is a matter of endowment and how much of it is due to a carefully designed policy? Who are the beneficiaries of Russia's newfound wealth and power? Using the chess metaphor, what is Putin's end game? These are some of the questions we will seek to address in what follows.

There is something about petroleum that is controversial and intriguing. And there is something about Russia that is mystifying and absorbing. When the two converge in a study of Russian petroleum, the result is bound to be tantalizing and engrossing, more like a fictional "who done it" than an accountant's annual report. The role of oil and gas in Russia is a tale of discovery, intrigue, corruption, wealth, misguidance, greed, patronage, nepotism, and power—except for the absence of sex (and who knows?), a little something for everyone. Admittedly it is a story that often bears considerable similarity with those of other oil-producing countries. Yet as with all things Russian, it has many features that are unique to Russia. Most significantly, after a long period of failure to sustain itself as a military superpower, Russia has emerged—even if inadvertently—as a different breed of superpower, one whose power rests on economics and energy. How is it using that clout and what does this imply in the years ahead? We begin in the beginning with a look at how Russia emerged as the world's largest producer of petroleum in 1898 and how what happened in the immediate years that followed has been repeated several times since.

1

Russia as an Early Energy Superpower

THE EARLY YEARS

Although they were unaware of its ultimate potential at the time, seventeenth- and eighteenth-century residents of what was to become Baku knew about and used the region's petroleum and natural gas. In fact, historians date the discovery of petroleum in the Baku area to a much earlier time. They point to the Parsees, a fire-worshipping cult that appeared centuries ago.[1] These followers of Zoroaster built a temple seven miles outside Baku that served as a holy site until 1880. Its perpetual flames were probably fed by natural gases escaping from the abundant deposits under the temple.[2] Even Marco Polo during his thirteenth-century travels noted that traders were very active in carrying oil-soaked sand to Baghdad.

Central Russian influence in Baku and the Caucasus in general came relatively late. After the fall of Constantinople, control of the Black Sea fell to the Turks, who kept the Russians out of the area for several centuries. On the other side of the Caucasus the Persians had control of the Caspian Sea. Ivan the Terrible pushed Moskovy's influence down the river Volga to Astrakhan on the north shore of the Caspian Sea in the sixteenth century, but formal Russian control of Baku did not come until the conquest of the area by Peter the Great in 1723. Once in command, Peter sought to ship some of the region's kerosene to St. Petersburg for possible use, but his advisers thought it was not worth the effort.[3] It did not matter much since shortly there-after, in 1735, the Persians regained Baku and impeded what little petroleum trade with the north there was. It was only in 1806 that the Russians recaptured Baku and in 1813 that they finally signed a peace treaty with Persia that authorized the transfer of control over most of the Caucasus from what is today Azerbaijan to Russia.[4]

Before the arrival of the Russians, petroleum extraction was very primitive. For centuries indigenous petroleum traders had to extract the petroleum with rags and buckets. By using hand labor they were able to increase the depth of some of the pits but it was all quite unso-phisticated. When the Russians came, they were able to improve the technology somewhat and production increased accordingly. In 1848 a Russian, A. F. Semenov, drilled the first relatively deep well, but even then the well was only sixty to ninety feet deep.[5]

In 1821, after their reconquest of the area, the Russians set up a special franchising system for those who wanted to produce and sell petroleum. Given how few hours of light there are in St. Petersburg during the winter, by 1862 there was a good market in the north for kerosene made from Baku's petroleum.[6] The rights to drill and pump petroleum on a specific site were extended on a monopoly basis by the state for four-year periods.[7] However, the lease could be revoked at the end of that time and there were no options for renewal. This deterred some investors and precluded more serious exploration and drilling activity, causing the leaseholders to extract as much as they could dur-ing the four years of their lease with little or no thought about maxi-mizing the long-run output of the area. This system prevailed until January 1873 when a more efficient public auction system was intro-duced.[8] As Table Intro.1 indicates, the changes facilitated a sharp increase in production. Although it was small to begin with, produc-tion doubled that year. The discovery of Baku's first gusher in June 1873 facilitated this growth.[9]

COMPETITION FOR MARKETS

These developments in turn attracted other prospectors, especially foreigners like the Swede Robert Nobel. Nobel arrived in Baku in March 1873 and soon came to exercise enormous influence in the area, not only as a producer but also as a refiner and marketer.[10] By 1883 production exceeded 1 million tons; by 1887 it exceeded 2 million tons (50 million tons equals approximately 1 million barrels a day). Equally significant, in 1877 and again in 1882, 1885, and 1891, strong and increasingly effective tariffs were passed that helped to curb Russian imports of American kerosene.[11]

However, it took more than tariffs to stem the flow. Russia had become an early battleground for oil producers seeking to carve out exclusive markets for themselves. John D. Rockefeller and his Standard Oil of the United States, as the world's largest producer, had taken over a dominant share of the Russian market. Eventually high tariffs on imported oil made it less attractive to import, but before Standard Oil and its U.S. petroleum could be pushed out of the Russian market, the Nobels—Robert and his brother Ludwig— had to find some way to facilitate the shipment of their petroleum and kerosene from Baku to the urban centers of Moscow and St. Petersburg. Just as during the Crimean War it was easier to move troops from Paris and London to the Crimea than from St. Petersburg and Moscow, so it was easier to move kerosene from the United States to St. Petersburg than it was from Baku. Seeking a way to facilitate the flow of Russian oil to the north, in 1878 Ludwig, designed a pipeline to carry crude oil from the well to their refinery and then on to the Caspian Sea. To carry large enough quantities across the Caspian Sea to make the venture profitable, he also conceived of and con- structed the first oil tanker, the *Zoroaster*.[12] His tankers docked at Astrakhan, where the oil was transferred to barges that then moved up the Volga. A storage depot was established in Tsaritsyn (later to become Stalingrad and now Volgograd) where by 1881 the petro- leum could be reloaded onto railroad cars, a convenience that was particularly important in the winter when the Upper Volga was fro- zen. One result of Nobel's innovation and the government's higher tariffs was the all but complete halt of kerosene imports from the United States. Imports, which were 4,400 tons in 1884, fell to 1,130 tons in 1885 and to an almost unnoticeable 22 tons in 1896[13]—one loss (temporarily) for the Rockefellers.

The cultivation of domestic markets was followed by an effort to expand foreign markets. For obvious geographical reasons, Persia had always been a major consumer of Baku's oil. For equally obvious geo- graphical reasons—that is, Baku's location on the essentially landlocked Caspian Sea—it was difficult to supply other regions of the world, including St. Petersburg. Since this was before Stalin came along to build his canal network, the challenge at the time was to break through the barrier of the Caucasus Mountains to gain access to the Black and Mediterranean Seas and thus to the ocean routes beyond.

Only in 1878 when the Russians pushed the Turks out of Batumi on the Black Sea did a new route became a realistic possibility. Shortly thereafter, led by A. A. Bunge and S. S. Palashkovsky, a group of Russian oil producers in the Baku region obtained a franchise to build a railroad over the mountains from Baku through Tbilisi to Batumi. Since they were short of funds, they sought the help of the Nobels. Initially, the Nobels refused, fearing that their dominance of the Baku trade, especially their St. Petersburg markets, would be jeopardized by the additional competition. Not to be denied, Bunge and Palashkovsky turned instead to the French house of Rothschild. Having recently backed a refinery on the Adriatic, the Rothschilds were anxiously searching for a source of crude oil to free themselves from dependence on Rockefeller's Standard Oil.[14] The cork on Russian exports was pulled when that trans-Caucasian railroad was completed in 1883–1884. Table Intro.1 indicates how overall exports increased. Exports from Batumi, which totaled 3,300 tons in 1882, increased to 24,500 tons in 1883 and 65,000 tons in 1884, an amount equal to previous total exports from all Russian ports.[15] The flow soon became even greater when a forty-two-mile pipeline replaced the most rugged portion of the railroad route in 1889.[16]

RUSSIA AS AN OIL EXPORTER

The increased flow of Rothschild's petroleum from Batumi and Nobel oil via the Volga put competitive pressure on Standard Oil's markets in England. The era was one of oil abundance, and sellers vied to under-price their competitors. Angry over the threat presented by Russian oil to his English and European markets, Rockefeller and Standard Oil retaliated in what was to become a familiar pattern by cutting prices. For a time this tactic succeeded, but ultimately the Russian producers prevailed and carved out a share of the market for themselves. Whereas the combined Rothschild-Nobel share of the British market amounted to only 2 percent in 1884, by 1888 it had expanded to 30 percent.[17] Overall, however, compared to worldwide American exports, Russian exports were relatively more important only in Asia. Thus in 1897, 75 percent of American exports went to Europe and 16 percent to Asia, whereas only 59 percent of Russian exports went to Europe but 38 percent went to Asia.[18] The pattern was much the same in 1913.

For almost twenty years the petroleum flowed so readily in the Baku region that there seemed to be no reason to develop new fields or exercise much care in pumping existing fields. The waste was enormous, not to mention hazardous. In what was to become a standard reaction in the years to come, visitors were appalled by the inefficiency, sloppiness, and lack of care exercised by Russian petroleum operators.[19]

Still, little changed as long as the oil kept flowing. Moreover, the per capita consumption of oil—or more appropriately, petroleum products—was much lower in Russia than it was in other advanced countries in the world. This was due in large part to the lower standard of living at the time in Russia. In the late nineteenth century, for instance, Russian consumption of kerosene per capita was one-half of that in Germany.[20] And since domestic productive capacity exceeded domestic petroleum needs, petroleum producers generally sought to divert a portion of their output to foreign markets. For example, during the good production years of 1903 and 1904, the Russian-based producers exported 16 percent of their total production (see Table Intro.1). In 1904 absolute petroleum exports reached their peak of 1.8 million tons. However, because domestic consumption by that time had increased, the relative share of petroleum exports earlier in 1890 and 1892 was actually higher; then 22 percent of all petroleum produced was exported.

Not surprisingly, therefore, Russian petroleum exports often exceeded those of the United States during the late 1890s and the early twentieth century. And if they were not the largest exporter, the Russians were certainly the second largest. It is difficult to tell precisely which years the Russians out-exported the Americans because the data are incomplete. When export-import data for crude oil are available for comparison, the United States statistics on imports and exports of petroleum products begin only in 1920.[21] Net American exports of refined products were high, but oddly enough, American imports of crude oil from 1920 to 1924 were even higher.

Nevertheless, as Table Intro.1 indicates, Russian production exceeded American production from 1898 to 1902, and virtually all of it came from the wells around Baku. In that four-year window, Russia was the largest producer of petroleum in the world. The Middle East was viewed as a barren desert then, and it was not until 1938 that the ARAMCO Consortium discovered oil in Saudi Arabia. The only other oil-producing areas of note at the turn of the century were in the Dutch East Indies and Mexico. Even with that, in 1897 Russia and the United States accounted for about 95 percent of the world's production.[22]

The high point for Russia was in 1901, when Russian production reached a pre-revolutionary peak of 11.987 million tons. The comparable figure for the United States was 9.468 million tons (see Table Intro.1). But while American production of crude oil reached 12 million tons the following year and continued to climb every year but one until 1924, Russian production did not exceed the 1901 level until 1929. Why did Russian production decline after 1901?

Initially there were rumors that the fields of Baku were running dry. Such rumors were not easily dispelled. Indeed, one early effort to set the record straight was by the noted Russian scientist Dmitry Mendeleyev, who wrote a paper entitled "The supposed exhaustion of the Baku oilfields."[23] Output did decline but not in the country as a whole. The Russians sought to cope with the drop in output in existing wells in a variety of ways. First, as production fell in some of the older Baku fields, prospectors drilled new fields nearby. Second, new deposits outside the Baku region were discovered. Whereas Baku accounted for 96 percent of all Russian production in 1897, by 1910 it made up 85 percent and by 1913 even less.[24] New fields that opened up at Grozny in Chechnia, at Emba (300 miles to the north on the northern shore of the Caspian Sea), and at Maikop, only fifty miles from the Black Sea, accounted for most of the difference.

Although there is some reason to believe that the existence and even early production at some of these sites as at Grozny predated the arrival of the Nobels, many of the more important fields were subsequently developed by foreigners like the Nobels, especially with English capital.[25] Western assistance also helped improve the technology. Learning how to drill deeper produced the quickest results. The commonly used Russian drilling methods, which often relied on wooden, not metal, tools, made it difficult to go deeper than 300 feet. By the end of the nineteenth century, Nobel and some of the other foreign companies were drilling wells more than 1,800 feet deep. With the help of the American-produced rotary drilling system, by 1909 the wells went as deep as 2,400 feet.[26]

Yet ultimately the Russians could not prevent a sharp decline in their production and exports. The drop was partly due to the failure of Russian companies to import the necessary technology. In what will turn out to be a recurring pattern, few Russian companies bothered to keep up with the rapidly changing refining and drilling techniques. Russian oil companies also conformed to the period's dominant international trend: ruthless corporate scheming and bitter rivalries.

Inevitably, the jockeying for market share around the world by companies such as Standard Oil and Shell had some impact. Price cutting was a common tactic. As a result, many producers cut back on their operations and occasionally went bankrupt. Recurring depressions had the same winnowing impact. Russian markets were not immune to such rivalry. In 1911 Shell became a major player in the Russian market when it purchased the Rothschild family's petroleum holdings. Shell entities then produced 20 percent of Russia's petroleum output, second only to that pumped by the Nobels. Since the revolution and expropriation were only six years away, it was probably one of the smartest sales the Rothschilds ever made. But most deal making of this sort involved financial juggling, not technological innovation, and thus added little to the country's productive capabilities.

Also hurtful to Russian production was the Czarist government's decision in 1896 to change the concession system that had governed Russian oil production.[27] In an effort to collect more revenue, the government instituted a combined auction royalty system. (It presaged the system that Middle Eastern states would come to use in the 1950s and 1960s.) At the time, however, the royalties demanded by the Czarist government seemed exorbitant, sometimes reaching as high as 40 percent. With the wisdom of hindsight, today that older Russian system looks like a bargain for a foreign investor. But considering that it was roughly seventy-five years before anyone else imposed such seemingly confiscatory terms, concession holders within Russia opposed the change and reduced output.

Most damaging, however, was the growing labor and civil unrest that hit the Batumi and Baku areas. Led in part by Stalin, strikes occurred in the Batumi area as early as 1901–1902.[28] They were followed in July 1903 by an oil worker strike in Baku. Interspersed between almost annual strikes in 1904, 1905, and 1907 were the activities of the reactionary Black Hundreds, supporters of the Czar who often resorted to mob action. What the strikers did not pillage or burn, the Black Hundreds did. Nor did the complex racial mix of Tatars, Armenians, Jews, Russians, and Muslim Turks and Persians add to the tranquility of the region once tensions erupted. The climax came during the 1905 Russian Revolution. Two-thirds of all the oil wells were destroyed. As a result, overall production fell by more than 3 million tons and exports were cut in half.[29] Whereas Russia produced 31 percent of the world's petroleum output in 1904, by 1913, due to the labor unrest, Russia's share had fallen to 9 percent.[30] Neither production nor exports were to recover significantly until long after the 1917 Revolution.[31]

The rather disappointing years of production and export in the decade before the Revolution should not obscure the fact that the petroleum industry in pre-revolutionary Russia had an important role to play (see Table Intro 1). Not only did Russia produce more petroleum than any other country for a short period of time but there were also periods when petroleum contributed in a fairly important way to the country's export earnings. True, petroleum exports never came close to matching grain export earnings, which accounted for 50 to 70 percent of the country's export earnings from 1895 to 1914.[32] But except for timber, petroleum was often the largest nonagricultural export. In the peak years of 1900 and 1901, petroleum generated 7 percent of Russia's export earnings, a foretaste of the much greater role petroleum would play after the revolution.

THE REVOLUTION

The 1917 Bolshevik Revolution had an immediate impact on oil production. The unrest caused by the workers' demands for more control over managerial decision making caused output to fall from 10.8 million tons in 1916 to 8.8 million tons in 1917. In some cases, workers' committees were formed to superintend management. Naturally this interrupted production. The Bolsheviks declared formal confiscation the following year, on June 6, 1918, officially nationalizing the fields.[33] Then production fell to 4.1 million tons.

The path of recovery was erratic because the revolution was followed by a counterrevolution that was supported by various foreign companies. As we saw, one of the more notable aspects of the pre-revolutionary period of Russian oil development was the important role played by foreigners. Swedish, French, British, and even American investors and operators devoted large sums of money in an effort to gain control and increase production. With the exception of the Rothschilds who sold out earlier to Shell, the revolution meant a loss for most of them. The decade that followed was marked by the efforts and intrigue of many of the former foreign operators to out-maneuver the newly empowered Bolshevik rulers to regain or repatriate some of their money. Even with the help of foreign military intervention, most failed, but oil men have always been more venturesome and bigger risk takers than most of us.

The Turkish occupation of Baku in September 1918 provided the opening the old investors had been waiting for. Aware that the

Bolsheviks were distracted by unrest in the north, the British sent in an expeditionary military force and in November 1918 pushed out the Turks. The British apparently had hopes of setting up an independent state of Azerbaijan. This was not solely an anti-Bolshevik gesture, but also an anti-Russian step to protect Persia and block Russian access to British India. Much the same type of maneuvering took place after the Second World War, only in the late 1940s the Soviets attempted to reverse the process and extend their influence from Azerbaijan into the northern part of Persia/Iran.

With the 1918 British takeover and denationalization of the Baku oil fields, hopes in the European stock markets soared on the expectation that the weak Bolsheviks would never come back. Moving fast in hopes that it could establish a presence in the area where previously it had been weak, Standard Oil of New Jersey signed a contract in January 1919 with the independent government of Azerbaijan.[34] It paid one-third of a million dollars for drilling sites. The Nobels toyed with the idea of selling their shares to the Anglo-Persian Oil Company but quickly grabbed yet another offer from Standard Oil. A tentative agreement was signed on April 12, 1920. Despite the fact that the Bolsheviks retook the area later that month and nationalized the region's oil fields, Standard Oil remained convinced the Bolsheviks would not be able to hold on. Reflecting its confidence, it paid Nobel half a million dollars for some additional land. Ultimately Standard Oil paid Nobel several million dollars for its stock that had already become worthless. According to Robert Tolf, this Standard Oil purchase was later to constitute one-tenth of the entire American claim against the Bolsheviks for American property seized during the revolution.[35] However, this speculative fever was not limited to Standard Oil. Shell Oil, along with other European investors, also bought what turned out to be worthless shares.

While the Soviets had gained physical control over the territory, they soon discovered that without the technical and managerial help of foreigners and others who had fled the area they could not really operate the oil fields. Output continued to fall until it reached a low in 1921 of 3.781 million tons, a level not seen since 1889. To add to their headaches, the Bolsheviks also found that the Western oil companies had united to boycott Russian oil exports, a pattern that was fairly common whenever oil fields were nationalized (these boycotts were usually only partially successful), at least until the late 1960s.[36] Formed in mid-1922, the Front Uni represented an oil consortium of fifteen companies, all of which promised they would not buy Russian "illegally produced petroleum." But because of Western greed and Russian connivance, the

Front Uni's embargo was broken even before it began to operate. Shell, itself a leader of the boycott, made a purchase of Russian oil in February 1923 and the French followed soon after.[37]

THE FOREIGNERS RETURN

The oil company embargo broke apart even earlier, particularly after it looked like Lenin had come to recognize that the nationalization of private property and the expulsion of foreign companies was a mistake. Acknowledging that they could not properly operate their newly nationalized oil fields, the Soviets began to solicit foreign help and the oil companies responded. Lenin personally approved such measures under the New Economic Policy (NEP), which authorized extending concessions for foreigners. One of the first to respond to the Soviet request for help was an American company, the Barnsdall Corporation.[38] Signed in October 1921, the Barnsdall contract actually predates the embargo. This was an important breakthrough for the Soviet Union. Not only did Barnsdall help the USSR restore production but it also served to attract several other foreign companies, including British Petroleum, the Societa Minerere Italo Belge di Georgia, and eventually a Japanese group in Sakhalin.[39] Once a breach had been made, the embargo failed.

The foreigners did what they were supposed to do. They restored the oil fields and started new ones. Barnsdall brought in advanced rotary drills and deep well pumps. Production rapidly recovered, and although there is some uncertainty as to how much Barnsdall made out of the venture, by 1924 when it left the Soviet Union, production was back up to 7 million tons. Production continued to increase, as did foreign technical help. Besides work at the wells, foreign help included American, German, and British assistance in the building of a second pipeline from Baku to Batumi, the French supply of a Schlumberger well-logging process, and American (Standard Oil of New York), German, and British support for refinery construction.[40]

Once output had recovered, the Soviets began systematically to revoke their concessions. By December 1930 most of them had been closed out. Standard Oil, however, was allowed to retain its concession at the kerosene refinery built in Batumi until at least 1935 and the Japanese stayed on Sakhalin until 1944.[41] But ultimately all foreign concessions were terminated.

At first glance it might seem that expelling foreign private companies was merely a response to traditional communist doctrine. But from the perspective of the post-communist Putin-era presidency, expelling foreign companies once Russia's own companies have begun to prosper has become standard Russian practice. Certainly today Shell Oil and Exxon-Mobil would agree that the harassment they recently faced in 2006 designed to make them walk away from their several billion dollar operations off the island of Sakhalin is more a form of nationalist than communist pressure.

Soviet petroleum policy then, just like Russian policy today, is not consistent. Almost at the time the Soviets were closing down Standard Oil's concession, they issued a new series of contracts. The critic Anthony Sutton records how companies such as Badger, Universal Oil Products, and Lummus were called back to rebuild and reconstruct refineries.[42] Having been supported by wartime Lend Lease contracts, some of their work continued until 1945. With this help, production and exports rose rapidly. Because of the damages inflicted by the Germans in World War II, production fell from its 31 million tons record in 1940 to 22 million tons in 1946. But with foreign help, as Table Intro.1 shows, by 1949 they had established a new production record

Soviet Control Inside but Capitalist Outside

As production increased, so did the amount of administrative control emanating from Moscow. In the early days, however, there was more control in principle than in practice. Theoretically control over industry was centered in the Supreme Council of the National Economy (VSNKH), which was created shortly after the revolution in 1917. VSNKH in turn derived its power from the Council of People's Commissars (CPK).[43] The CPK (the forerunner of the Council of Ministers) also created the Chief Oil Committee (*Glavny Neftianoi Komitet*) under the VSNKH on May 17, 1918. But only a few months later the Turks and then the British pushed the Bolsheviks out of the Baku region.

Meaningful control by the Russian authorities had to wait until they sent the British and Turks home in the spring of 1920. Then, recognizing the communication problem between Moscow and the Caucasus, the Chief Oil Committee authorized the creation of three local operating trusts. Azneft, which apparently was the most efficient and aggressive of the three, took over control of the Baku region. Grozneft took over Grozny, and Embaneft took over the fields in the Emba area.[44] The three trusts in 1922 formed a commercial syndicate, Neftesyndikat (later succeeded by Soiuzneft) to handle exports and other foreign activities.[45] Neftesyndikat proved to be a very aggressive monopoly. It joined together in a fifty-fifty partnership in 1923 with the English firm Sale & Company to market oil in the United Kingdom.[46] Neftesyndikat reserved the right to buy out all Sale & Company shares in ten years. This first British company was followed by the second. This time the partnership was between Neftesyndikat and Royal Dutch Shell. The Soviets also entered an arrangement with Standard Oil of New York to market Russian oil in the Near and Far East. They made other deals with British-Mexican Petroleum, Asiatic Petroleum, and Bell Petrole.

Neftesyndikat kept expanding and set up the Russian Oil Products (ROP) company in London jointly with Arkos, a Soviet foreign trade organization set up by the Soviet Ministry of Foreign Trade. By 1925 Russian Oil Products had its own filling station network. The Soviets also set up a filling station network in Germany called Derop through its subsidiary Deutsche-Russische Naptha Company. Ultimately other wholesale and filling station subsidiaries were formed in Sweden, Spain, Portugal, and Persia. With such a network to supply, Soviet oil exports increased rapidly. Soviet oil exports surpassed the previous level in 1926–1927 even though the production level was not exceeded until two years later. Reclaiming and in some cases going beyond its pre-revolutionary penetration, Soviet oil had an important impact in world markets. According to W. Gurov, who at the time was chairman of Soiuznefteeksport, at their peak from 1929 to 1933, Soviet oil exports amounted to 24.8 million tons over the five-year period. This accounted for 17 percent of all the petroleum imported by West Europeans.[47] Soviet statistics also show sales to the United States of as much as 50,000 tons in the peak year 1930.[48]

The most important purchaser by physical volume and market share was Italy. According to Gurov's calculations, Soviet oil accounted for 48 percent of Italy's total oil imports during the ten-year period from 1925 to 1935. In addition to their economic significance, these exports took on political importance after Mussolini became prime minister in 1922 and dictator in 1925. In other words, politics, at least in this instance, was no barrier to export. Only in 1938 and 1940 did the Soviet Union refuse to export petroleum to Mussolini's fascist government.[49] The Soviets were only slightly more discreet in selling petroleum to Hitler's Germany. Sales remained at the relatively constant level of 400,000–500,000 tons until 1936.[50] Exports to Germany then fell to about 350,000 tons in 1936 and to 275,000 tons in 1937. In 1938 and 1939 they dropped to almost nothing but shot back up to 657,000 tons in 1940 after the signing of the Nazi-Soviet Pact. In fact, in 1940 Soviet sales to Nazi Germany accounted for 75 percent of all Soviet petroleum exports that year.

PETROLEUM AND THE SOVIET BALANCE OF TRADE

Given their magnitude, Soviet petroleum exports were important not only for the purchaser but also for the Soviet balance of trade. Whereas before the revolution petroleum exports at their peak accounted for 7 percent of Russia's export earnings, in 1932 Soviet petroleum earnings generated 18 percent of total Soviet export receipts. That was a pre–World War II record. Exports in that record year amounted to 6.1 million tons and accounted for 29 percent of total production. Soviet net exports of petroleum far exceeded American net exports in 1932–1933. It should be pointed out that while the 18 percent share of oil exports in overall Soviet export earnings was due in part to the increase in the physical volume of petroleum exports, it was also due to the sharp fall in Soviet grain exports. Forty years earlier, when Russia was the bread basket of Europe, grain exports accounted for 70 percent of national export earnings. However, by the twentieth century, Russia's role as a grain exporter had diminished so much that grain generated only 53 percent of the country's export earnings. Then with the advent of communism and collectivization in particular, grain never again accounted for as much as 22 percent of the export volume. On those rare occasions when the Soviets were able to export grain, these exports seldom amounted to more than 10 percent of the country's overall total export revenues.

Important as petroleum was, however, it was not as crucial as some observers thought. Sutton, for example, mistakenly asserts that petroleum exports "became a significant factor in Soviet economic recovery, generating about 20 percent of all exports by value; the largest single source of foreign exchange."[51] In fact, in 1928 petroleum accounted for only about 14 percent of all earnings.[52] Moreover, the relative earnings of timber exports exceeded those of petroleum throughout the 1920s and 1930s, often by a substantial margin. For that matter there were years such as 1922–1923, 1926–1927, 1930, 1931, 1937, 1938, and 1940 when grain earned more than petroleum despite poor harvests and widespread famine.

The fall-off in petroleum exports after 1932 was due to the depression that afflicted all exports. From a peak of 3.2 billion rubles in 1930, Soviet export revenues fell to 2.8 billion rubles in 1931 and kept falling yearly (except for 1937) until they reached a mere 462 million rubles in 1939. (All trade figures cited here are stated in terms of constant 1950 ruble prices.) Reversing the trend, export revenue rose briefly in 1940, but this was a by-product of the Nazi-Soviet Pact. Because of the increase in the sale of petroleum to Germany, exports to Germany amounted to 50 percent of all Soviet exports (including petroleum) that year. In part, some of the reason for the drop in exports was that the Soviet Union began to need more of its raw materials for its own domestic production needs. More important, Soviet efforts were undercut by the collapsing economies of their customers. Depression may be a capitalist disease and it may have had no ostensible effect on the internal workings of the Soviet economy, but there is no denying that a depression of this length and magnitude was bound to have a devastating effect on the world demand for raw materials. This in turn affected Soviet petroleum export earnings. The Soviets quickly realized that they were exporting more but earning less. Thus, while 4.7 million tons of petroleum exports earned them 548 million rubles in 1930, two years later in 1932 when they increased export volume to 6.1 million tons, they generated only 375 million rubles in revenue.

It would take more than twenty years before the volume of Soviet petroleum exports would exceed the 1932 level. While production continued to expand, at least until the chaos of World War II, Soviet authorities began to direct more and more of the country's production inward. This was partly because of the realization that exports could be sold only at a low price and partly because the growing Soviet economy came to need more and more petroleum at home. Soviet planners also had to deal with a drop in yield at the Baku oil fields. That had an adverse effect on the state's efforts to meet the yearly production plan targets. The annual plan was a system introduced by Joseph Stalin in 1927–1928 to stimulate economic growth. When the USSR nationalized all the country's factories and means of production, it also did away with the private profit and loss system. But the Soviets needed an incentive system, so in place of profit and loss the state set out yearly and five-year plans specified in physical terms such as meters, tons, and product units. Managers and workers were rewarded with bonuses when the plan targets were met and penalized when they were not, and it was a disappointment when oil production fell temporarily in 1932. Fortunately it rose again sharply in 1934 but thereafter increased only modestly. As a result, petroleum output lagged far behind the targets set out for the Second Five-Year Plan, which ended in 1938. Production totaled 30 million tons, significantly behind the 46.8 million goal.[53] With the technology then at their disposal, the Soviets could not increase the production rate at their traditional fields in Baku and at Grozny.

Yet just as production seemed to be tapering off in the Caucasus, important new fields were discovered in the region between the Volga and the Urals. Eventually called the "Second Baku," the first discoveries in this area were made as early as 1929.[54] As in earlier years, however, a shortage of proper drilling equipment delayed the region's expansion. Only after the Second World War was petroleum in the newly discovered fields produced in large quantities, and the region, particularly its giant field at Romashkino, then came to outproduce Baku.[55]

CONCLUSION

There was nothing unique about what happened to the Soviet petroleum industry prior to the Second World War. Many of the trends and practices had already been established in the pre-revolutionary years and, as we shall see, would be repeated after the Second World War ended. For that reason it is worth summarizing what happened so that in the pages ahead we can more easily note the similarities when they recur.

To sum up, foreign help was very important to the Russian petroleum industry prior to the revolution as well as before the Second World War. That includes technological assistance at the drilling, extracting, and refining stages. Nor did the Soviets refrain from seeking foreign help to facilitate the foreign marketing of their petroleum. Often that meant selling to companies like Standard Oil or Shell so they could do the distributing. In other instances, it meant joining together with a Western company to form a joint Russian-local venture not only to handle wholesale distribution overseas but also on occasion to operate retail filling stations abroad as well. The concept of trading with multinational and notorious capitalist enterprises or even creating their own multinational network evidently posed no ideological hurdle for the Soviets. Nor, for that matter, was politics much of a barrier. The Soviets abandoned their previous party line and agreed to sell their petroleum to Mussolini's fascists and Hitler's Nazis, even when decency, if not self-interest, should have precluded such action. The politics of ideology was seldom allowed to stand in the way of the principle of profit.

One justification for seeking foreign help was the Soviets' periodic fear that their reserves might run out and that they were utilizing their output ineffectively. They expressed the same fears prior to the revolution, and—as we shall see—this would recur in later eras. Increased production was essential, not only because of the need to supply domestic demand but also because of the role petroleum played as an earner of foreign currency. At its peak, in 1932, petroleum accounted for 18 percent of foreign earnings. That depression year also saw the Soviet Union export more petroleum than did the United States and probably more than anyone else in the world. But to export that much, the Soviet Union had to divert 29 percent of its crude oil production from domestic use within Russia, a level not reached again until 1976.

Petroleum is indeed important as an exportable commodity, but its importance depends not only on Russia's ability to pump oil but on prices in the world market. When crude oil prices fall, the impact on the whole Russian economy can be serious as it was in the 1930s and would be in the 1980s and 1990s. Conversely, when energy prices are high, Russia finds itself with unprecedented power. Prior to 1973, while it needed the earnings from petroleum exports to pay for its imports, the world still regarded the Soviet Union as a spoiler, a price discounter, willing if not eager to cut petroleum prices and unsettle the capitalist oil companies. After 1973, the Yom Kippur War, and the resulting Arab oil embargo, the Soviets switched tactics and, more often than not, they sought to sustain prices at a level as high as possible to enhance the country's earning power. Profits, not politics, became the priority.

2

World War II to 1987

Russia Looks Inward and Outward

To Hitler, Russia stood for wheat and petroleum, but Hitler's information was at least partially dated. Once under Soviet control Russia's grain surpluses diminished rapidly. Whereas pre-revolutionary Russia had exported 9 million tons of wheat in 1913, the most the Soviets could muster prior to the Second World War was 5 million tons in 1931, and to do that they had to starve their own people.[1]

But if the breadbasket of Europe was not as full as it once was, the oil wells were pumping and as attractive as ever. One of Hitler's highest priorities was to capture the Baku fields. Although German troops did not quite reach Baku, they did capture the Grozny fields in Chechnia in the north Caucasus. Even there, however, by the time the Soviet troops were forced to retreat, the oil fields were so badly damaged that Hitler was unable to derive much benefit from them. But in the process, Hitler did manage to deny their use to the Soviets. Moreover, the Germans disrupted supply routes from Baku to the north so that the Soviets had a hard time maintaining their fuel supply. The USSR was helped to some extent by Lend-Lease oil shipments of 2.7 million tons of petroleum from the United States. Nevertheless, by the time the war ended, many Soviet oil fields had been badly damaged, so that in 1946 Soviet oil production had fallen to 22 million tons, down 30 percent from the 1940 peak of 31 million tons.[2]

THE VOLGA-URAL REGION

To expedite the postwar reconstruction of both the fields and the refineries, the Soviet government swallowed its pride and once again sought foreign help. They also confiscated $1 million worth of oil field equipment from Romania as a form of war reparation.[3] Most of the Soviet effort was directed at reconstituting and expanding the traditional Baku area, but gradually they moved north toward the meagerly developed Volga-Ural region. Although exploration in the newer area predates the revolution, no oil was found there until 1929.[4] Even then not much happened, and when the Second World War started annual output was not quite 2 million tons a year. Some major finds were made in the Volga-Ural's Devonian geological level deposits during the war in 1944, but serious drilling work began only in 1955.[5] Despite considerable drilling effort, because of the wartime damage, output in Azerbaijan in the Soviet era, especially Baku, never fully recovered. Even in 1966, the postwar peak, Soviet oil producers were unable to equal the 22.2 million tons pumped in Azerbaijan in 1940. (After the collapse of the USSR, Western companies were brought in by the government of Azerbaijan and production soon surpassed earlier output.) Fortunately as more and more new fields were discovered in the Volga-Ural region, output there rose rapidly and that area soon outproduced Azerbaijan. As a result, by 1949, total output in the USSR surpassed the previous level of production (see Table 2.1).

Overall output in the Volga-Ural region continued to increase until about 1970. This included the field at Romashkino in the Tatar ASSR (Autonomous Soviet Socialist Republic). For some time this field was thought to hold the largest crude oil deposits in the world. But after 1965 the rate of increase in output per well in this region began to fall sharply.[6] The response was to seek some way to enhance "secondary recovery." As in many other parts of the world, the initial solution was to inject water into the wells to restore the pressure needed to facilitate the extraction of petroleum. But water injection was only partially successful. Occasionally it made matters worse. For a time the extra water worked and increased the petroleum yield from the well, but the Russians typically injected too much water; as a result, it often became more difficult than necessary to extract the petroleum the water was intended to flush out. Special pumps were required, and before long the workers often found themselves pumping out more water than oil.

While Soviet petroleum engineering was often very effective, there were other more advanced procedures that could have been used; but

TABLE 2.1 Petroleum Production (Crude)

Year	Price/barrel	USSR Barrels/day*	USSR Tons**	Russia Barrels/day	Russia Tons	USA Barrels/day	USA Tons	Saudi Arabia Barrels/day	Saudi Arabia Tons
1960		2943			119	7055		1315	
1965	11.20	4858	242.9		200	9014	427.7	2219	111.0
1966	10.86	5302	265.1		218	9579	454.5	2615	130.8
1967	10.57	5762	288.1		235	10219	484.2	2825	141.3
1968	10.15	6167	309.2			10600	502.9	3081	154.2
1969	9.63	6566	328.3			10828	511.4	3262	162.7
1970	9.09	7127	353.0		285	11297	533.5	3851	192.2
1971	10.86	7610	377.0			11156	525.9	4821	240.8
1972	11.64	8064	400.4			11185	527.9	6070	304.2
1973	14.52	8664	429.0			10946	514.7	7693	384.0
1974	46.07	9270	458.9			10461	491.4	8618	429.7
1975	42.04	9916	490.8		411	10008	469.8	7216	359.3
1976	44.11	10466	519.7			9736	458.0	8762	437.3
1977	45.04	11010	545.8			9863	462.8	9419	468.4
1978	42.15	11531	571.5			10274	484.2	8554	424.4
1979	85.39	11805	585.6			10136	477.0	9841	488.0
1980	87.65	12116	603.2		547	10170	480.2	10270	509.8
1981	77.46	12260	608.8		554	10181	478.8	10256	506.3

(continued)

TABLE 2.1 (continued)

Year	Price/barrel	USSR Barrels/day	USSR Tons	Russia Barrels/day	Russia Tons	USA Barrels/day	USA Tons	Saudi Arabia Barrels/day	Saudi Arabia Tons
1982	66.94	12330	612.6			10199	480.7	6961	340.2
1983	58.13	12403	616.3			10247	483.0	4951	240.3
1984	52.86	12297	612.7			10509	496.1	4534	219.0
1985	50.11	12040	596.7	10904	542.3	10580	498.7	3601	172.1
1986	25.63	12442	615.4	11306	561.2	10231	482.3	5208	252.6
1987	31.68	12655	625.2	11484	569.5	9944	467.3	4599	221.1
1988	24.71	12601	623.7	11444	568.8	9765	459.1	5720	276.5
1989	28.69	12298	607.2	11135	552.2	9159	429.0	5635	271.2
1990	35.62	11566	570.5	10405	515.9	8914	416.6	7105	342.6
1991	28.79			9326	461.9	9076	422.9	8820	428.4
1992	26.98			8038	398.8	8868	413.0	9098	442.4
1993	23.09			7173	354.9	8583	397.0	8962	432.8
1994	21.07			6419	317.6	8389	387.5	9084	437.3
1995	22.03			6288	310.8	8322	383.6	9127	438.5
1996	25.94			6114	302.9	8295	382.1	9265	446.3
1997	23.51			6227	307.4	8269	380.0	9481	455.2
1998	15.71			6169	304.3	8011	368.1	9544	457.7
1999	21.41			6178	304.8	7731	352.6	8911	426.2

Year	Price/barrel	Barrels/day	Tons	Barrels/day	Tons	Barrels/day	Tons
2000	32.88	6536	323.3	7733	352.6	9511	457.6
2001	27.34	7056	348.᠄	7669	349.2	9263	442.9
2002	27.36	7698	379.6	7626	346.9	8970	427.3
2003	30.62	8544	421.4	7400	338.4	10222	487.9
2004	39.57	9287	458.8	7228	329.2	10588	506.1
2005	54.52	9551	470.0	6830	310.2	11035	526.2
2006	65.14	9769	480.5	6871	311.8	10859	514.6

Key:

"Price/barrel" is crude price per barrel in 2005 dollars

* "Barrels/day" is crude oil production per day in the given country, in thousands of barrels

** "Tons" is crude oil production in millions of tons (for the given year)

Sources:

All figures from 2006 BP Statistical Review of World Energy

Accessible at BP Global, Reports and Publications, Statistical Review of World Energy 2006

www.bp.com/sectiongenericarticle.do?categoryId=9010943&contentId=7021566

Russia pre-1985 from Annual Goskomstat, Narodnoe Khoziastvo

because Soviet authorities restricted contact between their technicians and foreign specialists, many production methods used in the West were not familiar or were unavailable to Soviet petroleum technicians. Moreover, even when such knowledge was available, the Ministry of the Petroleum Industry lacked the authorization to import the necessary equipment. The purchase of foreign equipment had to be approved by Gosplan, the Central Planning Agency, and foreign currency for the purchase had to be set aside by the Ministry of Foreign Trade and the Ministry of Finance. Even then, because Soviet authorities did their best to prevent onsite visits by foreign specialists, the manufacturers of that equipment could not always demonstrate how to use it properly. Thus much of what the USSR imported at the time served no useful purpose. Moreover, during the Cold War, U.S. authorities did all they could to embargo the export of advanced equipment and technology that the Soviets needed for enhanced recovery.

WEST SIBERIA

When output per well in the Volga-Ural region also began to fall, the slack was taken up by the opening of new areas in West Siberia. In contrast to the long period from 1929, when the first oil was struck, until the late 1940s, when production in the Volga-Ural region finally began to reach a meaningful level, the lag between discovery and production in West Siberia was much shorter. Although exploration for liquid energy in the region began before the Second World War, the first find occurred by accident in West Siberia in September 1953.[7] A drilling team was delayed while sailing up the Ob River near the town of Berezov. On the spur of the moment they drilled a test well and found gas in what became the Berezovskoe gas field. It was seven more years, in 1960, however, before the first oil was discovered in a Jurassic zone near Shaim on the River Konda, a tributary of the Ob and Irtysh. The "super-giant" field in a Cretacean level at Samotlor, about 500 miles to the east, was discovered in 1965, and the first commercial-scale well was completed in April 1968.[8] Whereas it took almost twenty years for the Volga-Ural fields to move from discovery to delivery to consumer, it took only eight years in the West Siberian Tiumen region. By 1970 production had reached 31 million tons; by 1975 it was 145 million tons; and in 1977, about 210 million tons.[9]

Notice the pattern here. The output in West Siberia seemed to compensate for the drop in productivity in the Volga-Ural fields in the late

1960s and early 1970s, just as the output in the Volga-Ural regions, coming on line in the late 1940s and early 1950s, offset the declining output of Baku. So far, each time output in one major region slackened, the Soviets found a new region. It would be nice if Russian oil operators could continue in this leapfrog manner. The American geologist John Grace doubts that will happen, since most of the giant fields in Russia, at least those that are easily reachable in terms of production and transportation costs, have already been discovered. But this type of solution to their production problem also had a negative side. It postponed the time when the Soviets would have to face the need to use their resources more efficiently. This is important because outside analysts began to warn as early as 1977 that the Soviets would shortly run out of new fields to develop. In an April 1977 open report that made big headlines in the United States, the CIA predicted that without a substitute for water injection technology, Russia's annual output would drop sharply in the Volga-Ural region. As the CIA saw it, by 1985 Russian oil output would fall off so sharply that the USSR would no longer have enough petroleum to export. In fact, it predicted that by the mid-1980s, the USSR and its East European allies would be forced to import 3.5 to 4.5 million barrels a day (175–225 million tons). As we shall see, the CIA's predictions were wrong; none of that happened.

THE SOVIET PLANNING, PRODUCTION, AND INNOVATION SYSTEM

The Soviet planning and incentive system and the special peculiarities that affected the Soviet raw materials and petroleum industries all but guaranteed that the Soviets would have difficulty solving their efficiency and productivity problems. In fairness, it should be pointed out that an inability to manage innovation effectively was not an affliction brought on solely by the Russian Revolution. The revolution seems to have compounded the problem, but even before 1917, we saw when discussing drilling technology in Baku in the nineteenth and early twentieth centuries, Russia's existing methods lagged behind developments in the West. Invariably it was necessary either to import more advanced technology or bring in foreigners to run actual concessions.[10] Such lags were not necessarily characteristic of all pre-revolutionary and pre–Five Year Plan technology, but they were widespread enough to cause suspicion that something deeper than a poor incentive system was at fault. There are a number of explanations: Russia was too remote to be affected by the West European Renaissance, Napoleon never stayed in Russia long enough to bring with him the reforms of the French Revolution and their emphasis on scientific enlightenment and rationality, widespread literacy was lacking until midway into the twentieth century, and the oppressive legacy of authoritarian and rigid governments before and after the revolution discouraged initiative. A mixture of all these factors probably contributed to the problem. Whatever the exact explanation, there is no doubt that Russian culture and history combined with the Soviet system of central planning and a lack of economic incentives stifled creative thinking, at least in the economic and technology spheres.

The Soviet development of the turbo-drill illustrates how shortcomings in the Soviet process of producing, planning, and innovation affected the development of the Soviet petroleum industry. In a perceptive analysis, Robert Campbell of Indiana University explains why the innovation-shy Soviet petroleum engineers were nonetheless prodded into developing this unique process which could utilize lower quality steel pipe.[11] To drill effectively using rotary drilling, the driller must have good-quality pipe that can withstand increasing tension and pressure as the drilling goes deeper. With poor-quality steel pipe, breakdowns, cracked pipe, and tool-joint failures are endemic.[12] This means not only an increased need for replacement pipe but lost time spent on repairs and lifting and lowering the portions of the pipe string that remained intact. With the type of pipe available to the Soviets at the time, they could drill down only 2,000 meters.[13] In 1950 that was the depth of almost 90 percent of Soviet wells. Though inefficient, it was adequate for the drillers in the Baku region. The Soviets were able to satisfy their needs, albeit with a good deal of waste.

While shallow wells may have been suitable for Baku, they were of no use in the producing fields in the Volga-Urals region where oil and gas deposits were much deeper. Furthermore, the use of the traditional rotary drill process with low-tensile-strength pipe meant that the drillers could reach the 2,000 meter depth only when the ground was soft and the rock not too hard. But as Campbell pointed out, Soviet drill pipe at best was made from what in the United States is considered unsatisfactory grade D and some higher grade E steel.[14] In the United States, drillers restricted themselves to the use of grade E steel and pipe or even higher grades.

Why did the Soviet Union, as the world's largest producer of steel, not produce higher quality steel? As long as the Soviet system placed stress on quantity rather than quality of production, the Soviet manager had little or no incentive to produce the higher grade steel. His pay depended not on producing high quality but on producing as much quantity as possible, usually measured by the weight of the steel.[15] Generally, the Soviet manager was not concerned about whether his product was sought after in the marketplace. It was usually enough that his product was produced and transferred from the factory floor. Once that happened and a factory achieved its physical output plan targets, the workforce would then share in the enterprise bonuses.

To ensure that such bonuses would be forthcoming, the factory manager devoted virtually his whole effort to searching for ways to increase production. Over time, Soviet managers developed a fine-honed sense of just how this should be done. Those who did not succeed were discarded along the way. Soviet economists soon discovered, however, that a single-minded devotion to quantity led managers not only to ignore quality and variety but to dispense with them. Every time, for example, a machine was shut down to change size or to improve the process, less time was available for production. Occasionally a change in process might lead to faster or improved production, but there was always the possibility that the innovation would not succeed and production would not increase. By contrast, there was always the certainty that in switching production models, production would be curtailed at least temporarily. Because few managers were willing to take such risks, quality improvement inevitably suffered. In the Soviet system, innovation was disruptive and therefore to be avoided.[16]

For the oil drillers, this meant that the steel manufacturers they depended on had no incentive to produce or even contemplate producing the higher grade qualities of steel.[17] In his colorful way, Nikita Khrushchev put it vividly when he complained: "The production of steel is like a well-traveled road with deep ruts; here even blind horses will not turn off, because the wheels will break. Similarly, some officials have put on steel blinkers; they do everything as they were taught in their day. A material appears which is superior to steel that is cheaper, but they keep on shouting 'steel, steel, steel!' "[18] While this attack was delivered in the context of criticism of the planners' inability to switch to new, more sophisticated and innovative industries like electronics, computers, and chemicals, Khrushchev's complaint was equally valid when addressed to the need for qualitative improvements within the steel industry itself.

The planning system was equally ill-suited for locating new deposits. Since planning targets were usually spelled out in terms of some physical measure, for those in agencies like the Ministry of Geology whose work involved drilling, the most reasonable index seemed to be the number of meters drilled. Supposedly the more meters drilled, the better the performance. But Soviet geologists soon discovered that the deeper they drilled the longer it took them and the less drilling they did.[19] As a result, the geologists quickly developed the practice of drilling shallow holes. As an article in *Pravda* pointed out, "Deep drilling means reducing the speed of the work and reducing the group's bonuses."[20] A description of the area sounded more like a smallpox rather than a mining report. "In some places, the land is becoming increasingly pitted with shallow, exploratory holes drilled in incessant pursuit of a larger number of total meters drilled." It was not surprising, therefore, that "there are geological expeditions in the Republic of Kazakhstan that have not discovered a valuable deposit for many years, but are counted among the successful expeditions, because they fulfill their assignment in terms of meters. The groups that conscientiously 'turn up' deposits are often financial losers."

Moreover, even if the drillers from the Ministry of Geology found a field, they bore no responsibility for determining its size. Consequently, the actual producing ministries also had to maintain their own drilling units. In some instances there were two and on occasion as many as three separate drilling agencies duplicating one another's work.[21] Undoubtedly it would have been much more efficient to base the drilling team's compensation on the amount of raw materials actually recovered, but this was resisted by the planning agencies and the Ministry of Geology, which feared such a shift would disrupt its planning procedures. As often happened, they had confused the means—that is, how many meters are drilled—with the end, how much oil was found. Another way the Soviet planning system created institutional blockages to a more efficient utilization of mineral deposits was that responsibility for production and drilling was usually divided up among several large ministries or state committees such as the Ministry of the Chemical Industry, the Ministry of the Gas Industry, and the Ministry of the Petroleum Industry.

Unfortunately, nature did not always break itself up into the same neat and precisely defined categories. This also helps to explain why so much natural gas was burned off, that is, flared. Virtually none of the flaring was done by enterprises within the Ministry of the Gas Industry. Most of it was done by drilling units working within the Ministry of the Petroleum Industry. They produced the gas as a by-product in extracting petroleum, which, after all, was their main concern.[22] Therefore, the plan fulfillment efforts of the Ministry of the Petroleum Industry set out in tons of petroleum produced were little affected by what happened to the by-product, natural gas (a by-product of petroleum extraction). Since Petroleum Ministry officials received no credit for producing gas, they did not concern themselves with building gas pipelines to move the gas to market. Why should they bother? To rid themselves of the nuisance, more often than not they simply flared it.

THE CIA'S PREDICTION OF A SHARP DROP
IN PETROLEUM PRODUCTION

Given so many counterproductive and illogical practices, it is easy to see how CIA analysts could conclude that despite the Soviet Union's large land mass, Soviet petroleum output would drop sharply. Since the oil field operators would most likely continue to flood more and more of their best oil wells, it seemed inevitable that before long the USSR was bound to become a net petroleum importer. Moreover, were production to continue to fall as the CIA predicted, the USSR was sure to find itself with problems that extended far beyond the Ministry of Petroleum. Petroleum was virtually their only hard currency export, and if they could not export it they would not be able to earn the hard currency they needed to pay for imports. Were that to happen, they would be unable to fund the $6.5–8 billion a year they periodically spent on meat and grain imports.[23] As it was, even with the petroleum exports they frequently ended up with a trade deficit.[24] In 1975 and again in 1981, for example, the Soviet trade deficit exceeded $4 billion. It would have become even higher if the Soviets had been unable to respond by increasing their petroleum exports.[25]

RUSSIAN PETROLEUM AS DIPLOMATIC WEAPON

The CIA rightly concerned itself with the Soviet Union's ability to export petroleum. The earlier surge in Soviet petroleum output and the corresponding increase in exports in the 1960s and 1970s provided Soviet leaders with a particularly effective economic and foreign policy weapon. It opened doors in the third world for Soviet ideology and diplomatic initiatives that otherwise might have remained closed or just half open. Countries in the struggling regions of Asia, Africa, and Latin America in that era generally welcomed the radical rhetoric propagated by the USSR but often hesitated to turn their backs

completely on their former colonial masters for fear of economic repri-
sals and export embargoes. In particular, radical leaders in the third
world feared that if they became too tied to the Soviet Union or went
so far as to nationalize the Western oil companies' distribution network as did Cuba and Ceylon (renamed Sri Lanka in 1972), the United
States would arrange with the so-called Seven Sisters capitalist oil
companies to embargo the delivery of the petroleum essential to running their economies.[26]

First organized in 1928, the original Seven Sisters consisted of Royal
Dutch Shell, Standard Oil of New Jersey (today's Exxon), and the Anglo
Persian Oil Company (today's BP). After the breakup of Standard Oil,
the original three members were expanded to include Standard Oil of
California (today's Chevron), Standard Oil of New York (Socony
Vaccum and later Mobil Oil, now part of Exxon), the Texas Company
(what became Texaco and then Chevron), and what was once Gulf Oil
(now also Chevron). In contrast to today's world where we worry about
energy shortages, the reason the Seven Sisters joined together was to
deal with an overabundance of petroleum on the market and the price
cutting that resulted. Their purpose was to form a cartel and limit production and price cutting, and in the 1950s one of their main concerns
was how to deal with the Soviet practice of price cutting. They also
used their control of petroleum exports to punish third world countries
that nationalized properties owned by Western investors or otherwise
impinged on Western prerogatives.[27]

Man does not live by bread or oil alone, but the leaders in the third
world quickly discovered that it helps to have a little of both. Once the
USSR began offering to underwrite the growing ranks of rebellious
colonies with Soviet petroleum, this reduced the retaliatory powers of
the mother countries and the Western oil cartels, which often as not did
their bidding. Even when there was no formal embargo on petroleum
sales, such offers from Soviet officials were very much appreciated
because most of these former colonies lacked sufficient hard (convertible) currency for their needed purchases in the traditional energy markets. The Soviets in almost all cases were happy to sell oil at a lower
price (sometimes less than a dollar a barrel) or lend or barter their oil
without insisting on hard currency payments as a way to gain influence.
This was an important form of economic support for the East European
Communist and other Council of Mutual Economic Assistance countries (CMEA but more commonly referred to as COMECON) as well
as Cuba and most of the former African and Asian colonies including
India, Ceylon, Pakistan, Guinea, and Ghana.[28]

Since the Soviet Ministry of Petroleum was an instrument of the state, there was little resistance within the Soviet Union to using the country's petroleum this way. The first priority was to provide for domestic needs. The next was to use petroleum exports to generate the money needed to pay for the Soviet Union's and Eastern Europe's hard currency imports from the capitalist world. Anything extra available for export could then be used to promote the state's political goals. Unlike a private petroleum company, the Ministry of Petroleum did not feel constrained by normal corporate profit and loss considerations. To say the least, profit maximization was not an overriding objective. In fact, it usually played no role at all. Moreover, just as in the Czarist era, the richer countries in the outside world, and especially the Seven Sisters, were not particularly welcoming to exports of Soviet petroleum. Both Western governments and businesses remained unforgiving about the nationalization of the Baku oil fields. For example, until 1971 the British went so far as to prohibit oil dealers, including a network of as many as 400 Soviet-owned service stations located in the United Kingdom, from importing Soviet crude oil.[29] The Soviet response was to arrange for their English subsidiary to import crude oil instead from Finland, which was strange, since Finland is not often thought of as a petroleum powerhouse. In fact, Finland imported more than three-quarters of its petroleum from the USSR. Admittedly this was a pain in the neck and added an extra step for the Soviet Ministry of Foreign Trade officials in charge of their British subsidiary. But it also shows how difficult it was to exclude Soviet petroleum exports from world markets.

Such efforts to keep Soviet petroleum out of world markets was largely a result of the Seven Sisters' neither needing nor desiring to buy any petroleum from the Soviet Union. These companies regarded the Soviet Union as a spoiler and a disruptive influence—often accusing Soiuznefteexport, the Soviet Foreign Trade Organization responsible for exporting the country's petroleum, of dumping its products to force down the Sisters' petroleum prices and profits. Because there was pressure to keep them out of world markets, for some time Soviet petroleum export officials were relegated to dealing with impecunious and marginal consumers or working out under-the-table transactions.

As world energy demand grew, however, attempts to exclude the Soviet Union from the capitalist world's petroleum markets became harder. Even more important, by the mid-1970s it made increasingly less sense. The first serious Soviet challenge to important Seven Sisters markets occurred in 1960. ENI, an Italian energy company headed by

Enrico Mattei, had been attempting to break into the Sisters' club. In 1957, he signed a contract with Iran to buy petroleum from that country at concessionary prices that were higher than those that had been offered Iran by the Seven Sisters.[30] At the same time he offered to sell potential customers that Iranian petroleum at a cut-rate price. Increasing the pressure, Mattei broke ranks again in 1960 by arranging for yet another out-of-order petroleum purchase, this time a cut-rate purchase from Soiuznefeexport. As he did with Iranian petroleum, he sought to sell this petroleum to Seven Sisters customers in Europe by undercutting the prevailing Seven Sisters price. This was considered a direct challenge to the ability of the Seven Sisters to control prices and a serious destabilizing threat. In some quarters, the inability of the Seven Sisters to prevent low-priced Soviet petroleum from breeching their monopoly control was viewed as the end of the Seven Sisters monopoly.

But Soviet petroleum exports served as an equal opportunity spoiler. They not only undermined the Seven Sisters and their price control efforts but they also undercut the efforts of the Organization of Petroleum Exporting Countries (OPEC) member countries that were trying to do the same thing. Created in September 1960, OPEC was set up to prevent private oil companies from cutting the price of the petroleum they purchased from Saudi Arabia, Kuwait, Iraq, Iran, and Venezuela. These original members of OPEC attempted to do this by regulating how much petroleum each of these countries could produce and by doing so reduce worldwide supplies.

But while most of the world's major petroleum exporters were curbing their production and exports, the Soviets were expanding theirs. By 1975, they had become the world's largest producer of petroleum, overtaking the United States, which had maintained that distinction continuously since 1902 when it outproduced the largest producer at that time, Czarist Russia.

By refusing to go along with OPEC, the Soviets increased their political leverage as well as their earning power. For that reason, the 1973 OPEC oil embargo imposed on the United States and several European countries provided the Soviet Union with a golden opportunity. The tightening of petroleum markets that resulted from that OPEC embargo more or less brought an end to the USSR's bad boy image. The Soviet Union may have been a rogue, but OPEC members were no better, and in 1973, at least, were much worse. Thus after 1973, energy consumers around the world came face to face with the realization that reliance on energy supplies from the Middle East involved enormous risks. How much more risky could reliance on the USSR be?

As their industrial output continued to grow, Soviet leaders also sought to export some of their growing output of natural gas. At the same time, after 1973, whatever resistance potential customers may have had to buying petroleum and gas from the Soviet Union all but disappeared. Chastened by the 1973 Arab embargo, customers in Western Europe in particular began to search for ways to reduce their dependence on the now uncertain imports from the Middle East. The Germans were especially eager to gain access to other sources, and the Soviets could ship them natural gas—a cleaner fuel than oil—via an overland pipeline. Most of all, it was reassuring that petroleum and gas from the USSR would be unaffected by OPEC embargoes or sea blockades. To top it off, because of their outsider status, the Soviets were usually willing to undercut market prices.

In time, major net importers of energy such as the United States, Western Europe, China, and India came to realize that it was in their interest to encourage as much energy production in the world from as many different producers as possible. This obviously included Russia— the more supplies there were the better. This was particularly important for consumers of petroleum. If one supplier decided to withhold deliveries of its petroleum, importers could readily substitute with petroleum from another supplier. That reduced—but did not eliminate—the chance of a political embargo by an OPEC-type organization against a single consumer or group of consumers. But those who initiated such an embargo had to win support from like-minded exporters and even then there were bound to be some exporters such as Russia that would refuse to join in. There is always the danger that those exporters would use the opportunity to poach on others' customers and sign up new sales.

But while buying petroleum and natural gas from Russia has its advantages, there can also be serious risks, especially for consumers of Russia's natural gas. Because pipelines needed to supply natural gas are very expensive to construct, no one can afford to build a second standby pipeline from some other supplier as a reserve for emergencies. Thus even though the European pipeline network links up three major sources of supply—Russia, the North Sea, and Algeria—consumers of natural gas tend to become dependent on a single dominant supply source. This makes them vulnerable to the whims of that supplier. While LNG (liquefied natural gas), which can be delivered by seagoing tankers, could serve as a backup, it too requires billions of dollars in investment, not only for the special tankers that transport it but also for the expensive processing plants at the export site that freeze it and the plants at the import destination that return it to gaseous form. As a result, no one is willing or able to sell or buy LNG without an expensive infrastructure already in place, which explains why it is so hard to create a spot market for LNG, a market where buyers and sellers can agree to a sale at the last minute on the spot. As a result, unlike petroleum imports, which can be sent by tanker from any number of petroleum producers, if something happens to that natural gas pipeline, there are rarely any alternative natural gas supplies available to pipe in as a substitute.

The American President Ronald Reagan understood the political implications of all this and decided to do what he could to prevent the USSR from building a gas pipeline to Western Europe. He worried that if the Europeans became increasingly dependent on such supplies, as strong economically as Western Europe might be, they would soon find themselves vulnerable to Soviet political pressure. Reagan worried that as West European households and industries began to rely on Soviet natural gas, they would likely begin to think twice about countering Soviet political demands.

In an effort to deny the USSR such a weapon, Reagan launched an intense effort to prevent the pipeline's construction. In 1984, he asked his friend, British Prime Minister Margaret Thatcher, to prevent the English firm, John Brown Engineering, from selling the Soviets the compressors they needed to move the gas through the pipeline from the super giant Urengoi natural gas field in West Siberia to Germany. Similar pressure was put on General Electric, another manufacturer of turbines and compressors. These efforts failed, however, and the pipeline was eventually completed.

Once the pipeline was completed in 1985, consumers in Western Europe became quite comfortable importing Russian natural gas and using it in their homes and factories. While cold weather caused occasional delivery problems, the Cold War never did. As a good salesman, Viktor Chernomyrdin, when he was Minister of the Soviet Gas Industry, always insisted that he would never think of cutting off the flow of gas for political reasons. He continued to issue such assurances when the Ministry, in August 1989, was transformed into a hybrid state corporate entity which he called Gazprom. He abandoned the title of minister and called himself the CEO of the now newly created joint stock corporation. The promise that the ministry and then Gazprom would honor its contracts has been echoed by numerous other senior government officials including Chernomyrdin's immediate successor, Rem

Vyakhirev, who as CEO, sold much of the company's stock to nonstate entities in November 1992. Others frequently made the same claim. For example, in 2006, Igor Shuvalov, President Vladimir Putin's economic adviser insisted that "Europe will never have a more reliable supplier than Russia."[31] Or listen to Putin himself. At the Balkan Energy Cooperative Summit in Zagreb in June 2007, he insisted that "for four decades now, despite the serious and truly global changes in the world, Russia has never broken a single one of its contractual commitments."

Such assertions, however, overlook the fact that the Soviet Union in its day and Russia after 1991 have frequently terminated the shipment of energy supplies when a customer chose to oppose Soviet or Russian political or economic objectives. Yugoslavia under Tito, Israel in 1956, Finland in 1958, China in 1959, Latvia in 1990, Lithuania in 1990 and 2006, and Estonia in 2007 had their petroleum deliveries cut off. Later, Putin's regime halted or reduced the flow of natural gas supplies to Ukraine, Belarus, Georgia, Moldova, and even Bosnia. What passes as "the rule of law" in other societies became "the law of the rulers" under Putin. A contract commitment with state-controlled enterprises in Russia has never been a guarantee of performance nor a deterrent to arbitrary behavior by Russian entities. That was true in the Soviet era and it again became common under Putin. Concessions made at a time when Russia is weak and prices are low are invariably invalidated once prices rise again and Russia regains its strength. Put simply, higher prices increase Russia's bargaining power. Precedent is no guarantee that the Russians will not some day mend their ways, but it does suggest that President Reagan had legitimate concerns.

WILLIAM CASEY: DID HE PRECIPITATE THE COLLAPSE OF THE USSR?

Recognizing how important petroleum and natural gas production and delivery were to the Soviet Union's domestic and foreign well-being and influence, William Casey, appointed by President Reagan in 1981 to head the CIA, decided that the best way to undermine the USSR was to undertake an effort to cripple its energy sector. Given that four years earlier the CIA had predicted there would be a sharp drop in production, which would turn Russia from an oil exporter to an oil importer, it should be relatively easy to expedite that drop in oil production.

Fortunately for the USSR, neither it nor Russia became net importers—far from it. So was the CIA wrong? Production did peak in

1987 at 625 million tons and it did fall to 571 million tons in 1990 (see Table 2.1). Yet according to the CIA, the USSR and Eastern Europe should have been importing 175 to 225 million tons by then. But that did not happen. The USSR remained a major petroleum exporter until it disintegrated. After that, production in Russia itself did indeed fall sharply in the mid-1990s, but as we shall see in Chapter 4, this was because petroleum prices were low and taxes were high so the new private owners concluded they could make more money by stripping assets than by producing petroleum. These were not the reasons anticipated by the CIA.

The CIA prediction was far off the mark. In fact, in 2006 Russia again became the world's largest producer of petroleum. Nonetheless, the production and central planning problems on which the CIA based its 1977 analysis were real. The Soviets continued to inject too much water into the oil fields and the bureaucratic and central planning practices that characterized the Soviet economic system resulted in enormous waste and lost opportunities.

While the CIA devoted considerable effort to research and analysis of the problems that confronted the Soviet petroleum industry and its exports, once William Casey took over as the head of the CIA, he began to tackle the issue more aggressively. Some, including Peter Schweizer of the Hoover Institute and Yegor Gaidar, former Acting Prime Minister of Russia in the early years of Boris Yeltsin's presidency, have advanced the view that Casey sought to cripple the Soviet petroleum industry's export-earning capabilities to prevent it from generating the hard currency Russia so desperately needed to pay for its food and technology imports.[32]

Schweizer goes so far as to argue that the CIA under William Casey launched a complicated scheme that ultimately led to the collapse of the Soviet Union. As Schweizer tells the story, CIA chief Casey received authorization from his boss, President Reagan, to work with Saudi Arabia to weaken the Soviet petroleum industry. This was typical of Casey's out-of-the-box thinking. As he saw it, Casey reasoned that the Saudis would cooperate because they were angered by the Soviet invasion of Afghanistan, a brother Islamic country. Saudi Arabia at the time actively supported the Afghan guerillas fighting the Soviet occupiers. Before long the Soviet Union found itself bogged down there. Casey also sought to weaken the USSR at home.

From his own background in international finance Casey understood that the Soviet Union depended heavily on petroleum exports to pay its international bills. This included not only payment for massive imports of grain (by the late 1970s, the Soviet Union had become the world's largest importer of grain) but for imported factories (for example, large chemical plants) and technology that the USSR was unable to produce itself. These imports were also important in providing the Soviet Union with the wherewithal needed for its military-industrial complex. According to Schweizer, Casey thought that if he could somehow shrink the value of the USSR's petroleum exports, that shrinkage would force the Soviets to curtail their involvement not only in Afghanistan but elsewhere in the world. All of this suggests, however, that Casey was unconvinced by the earlier CIA predictions that Soviet oil output would fall. If those earlier conclusions had been right, the Soviet hard currency earnings would have been reduced without any need to seek Saudi help and, short of money, the Soviet Union would have been forced to withdraw from Afghanistan. A drop in hard currency export earnings would also have hurt industrial investment within the USSR itself. Except for the revenue earned from petroleum and to a lesser extent natural gas exports, the Soviets had virtually no other way to pay their external bills. Consequently, Casey sought ways to reduce the USSR's earnings from its petroleum exports. While he might not precipitate the Soviet Union's collapse, at least he could weaken its structure.

To implement this ingenious scheme, Casey sought out the Saudi leadership in 1985 and, according to Schweizer, urged them to increase their output and export of crude oil. By expanding world supply they would precipitate a drop in world oil prices. Casey argued that this would not only help the U.S. economy but would seriously hamstring the Soviet economy and presumably force the Soviets to curb their adventures in Afghanistan.

What was the exact cause and what was the effect even now is not known precisely. As reflected in Table 2.1, Saudi output fell to a sixteen-year low in 1985 after hitting an all-time high in 1980. Then after King Fahd's visit to Washington to see President Reagan in February 1985, the Saudis did pump more oil.[33] Output in 1986 rose 45 percent over 1985 (see Table 2.1).[34] But perhaps equally if not more important, increased petroleum pumped from the North Sea and West Siberia hit the market at the same time. As anticipated, average prices in 1986 fell to half of what they had been the year before, to $25.63 a barrel (see Table 2.1). By 1988, average prices dropped even further to $24.71.

We could only guess at the time what the impact of the fall in prices was on the Kremlin leadership. With the benefit of hindsight, Casey appears to have anticipated correctly. Relying on Politburo archives,

Yegor Gaidar reports that Soviet leaders were in near panic. The drop in prices, he says, cost Russia $20 billion a year.[35] Their financial condition was evidently much worse than those on the outside knew. It was widely believed that even if oil prices were to fall, the Soviets could use their large stocks of gold to pay their bills. But Yegor Gaidar now reveals that by early 1986, they had only $7.6 billion left, not the $36 billion in gold that most outside observers at the time assumed. Most of their gold had already been sold to pay for earlier grain imports. In 1963, for example, Khrushchev spent one-third of the country's gold to import 12 million tons of grain.[36] Once oil prices started to fall, not only did each barrel of petroleum exported earn fewer dollars but the drop in export earnings also forced the Soviets to reduce their industrial imports and the investment they needed to sustain oil production.[37] This in turn affected morale already shaken by the turmoil precipitated by Gorbachev's 1985 perestroika campaign. As a result, crude oil output began to drop sharply. By 1990, crude oil output was down about 10 percent (see Table 2.1), which meant a further reduction in imports and the need to borrow even more money from foreign banks and governments.

Because Gorbachev and his programs were so popular in the West, there were many calls to be supportive. This gave birth to "a grand bargain" proposed by Graham Allison, dean of the Kennedy School of Government at Harvard University.[38] But a growing number of foreign suppliers and bankers came to realize that the USSR's financial plight was so serious that the Soviets might not be able to repay any such loan. This in turn led them to withhold credits. This only served to increase anxiety in the Kremlin.[39] Yegor Gaidar recounts that as the financial situation continued to deteriorate, out of desperation Gorbachev found it necessary to contact Chancellor Helmut Kohl of Germany. He begged for immediate help, explaining that the situation in the USSR had become "catastrophic."[40]

All of this had a destabilizing impact on the USSR. By 1988, faced with intermittent bad harvests, an empty treasury, an increasingly unpopular war in Afghanistan, and a domestic economy in turmoil as it sought to free itself from some of the excesses of central planning, Gorbachev and some of the other Soviet leaders were finally forced to acknowledge that the Soviet Union had overextended itself.[41] Its economic wherewithal could no longer support its imperial pretensions. That explains at least in part Gorbachev's decision to begin the withdrawal of Soviet troops from Afghanistan on February 15, 1989. (It is hard to resist making comparisons with the United States fighting in Vietnam and Iraq.)

Admirers of CIA chief Casey credit him and his efforts with Saudi Arabia for forcing the Soviet Union's retreat in Afghanistan and by extension for the collapse of the USSR itself two and a half years later.[42] Undoubtedly the increase in world petroleum output and the resulting drop in price that followed seriously undermined the Soviet Union's international financial creditworthiness and its ability to support its own and its East European satellites' economies.[43]

But was the cause and effect so straightforward and so simple? Prices in 1985 did indeed fall from $50 a barrel (in 2005 prices) to $24 a barrel in 1988. While the Saudis did increase production in 1984, 1985, and 1986, they actually reduced production in 1987. Belatedly in 1988 they again made an increase but to a level less than they produced in 1980 and 1981. Whatever the cause, the Soviets did evacuate Afghanistan on February 15, 1989, but only after Soviet output hit its peak. The Saudis boosted output in 1990 by 70 million tons, much more than the 50-million-ton increase in 1988. But by 1990 the war had already come to an end. If the Saudi increase in production had such an impact on USSR prices, why didn't oil prices fall in 1980 when the Saudis were pumping two and three times as much oil as they pumped in the mid-1980s, and why did Saudi Arabia wait until the 1990s, rather than in 1985, after Casey's intervention, to make major increases in production?

As Gaidar's research into the minutes and correspondence of the Politburo makes clear, there is no doubt that the fall in world petroleum prices did hurt the Soviet Union.[44] But the collapse was due to more than the drop in oil prices. After 1987 there was also a drop in Soviet oil production, which also hurt earning power. The lower prices undoubtedly did contribute some to the drop in output, just as it was to do in the early and mid-1990s. But did lower prices have that much effect on the Ministry of Petroleum and its affiliates? In the Soviet era, profits and prices were not all that important as a stimulus to production. What mattered were targets set by the plan. Market incentives came into play only after privatization. Admittedly the increased use of water injection cited by the CIA in its 1977 report did hamper production, but it was not an unsolvable problem. The CIA prediction that Soviet oil production would fall and the USSR would soon become a major importer notwithstanding, we will see to the contrary in Chapter 5 that Soviet oil production did increase again—and substantially— after 1999.

While there may have been a connection between increased Saudi oil output, lower oil prices, and the Soviet Union's collapse, the Bill

Casey intrigue does not explain why the USSR did not collapse in 1980–1981 when Saudi output was at a record high, double what it was in 1986. Note even at the 1980–1981 high point of production, the USSR still produced more petroleum than Saudi Arabia. It was only in 1992 that Saudi output exceeded Russian output. Despite the elevated level of production, oil prices were actually at a record high then. It may have been that the anxiety created by the Soviet invasion of Afghanistan in 1980 had a greater impact on oil prices than the increased Saudi output. But the fact that prices did not fall until 1981— and that the Soviet Union was unaffected, at least in the early 1980s— suggests that while Casey's efforts to undermine the Soviet Union's economy may have had an impact, it cannot be argued that his conspiring with the Saudis was the sole or even the most important cause of its collapse. Nonetheless, Casey's involvement is yet another bizarre episode in this fascinating and ongoing interplay of geology, economics, ideology, politics, and greed.

3

Pirates Unleashed

Privatization in the Post-Soviet Era

THE USSR IS NO MORE

The disintegration of the Soviet Union unleashed a cascade of centrifugal forces, both political and economic. In 1992, after the USSR broke up into fifteen independent and occasionally hostile countries, a Moscovite traveling to Kiev or Minsk could do so only if he had a passport for foreign travel. If that Moscovite tried to ship goods to Ukraine, Belarus, or Uzbekistan, he would have to send them through customs, pay a tariff, and accept payment for his goods in something other than rubles. None of this had been necessary before when they were all brother republics within the USSR. Equally disquieting, Boris Yeltsin, the hero in putting down the August 1991 coup attempt and the duly elected president of Russia, had serious drinking and health problems (both physical and mental). This made it impossible for him to focus properly on matters of state. Yeltsin had no problem forcing the breakup of the USSR and spinning off the other fourteen republics (including Ukraine and Belarus that were Slavic), which before the revolution were provinces of the Russian empire. But in an action that haunts Russia today, Yeltsin decided that Russia would not let anyone else split off from Russia and so ordered his troops to put down an insurrection in Chechnia, a relatively unimportant but problematic region within Russia's boundaries. Unlike Ukraine and Belarus, which share Slavic ties to Russia, Chechnia is a Moslem rather than a traditionally Slavic region. It was forced into the Russian empire in the late nineteenth century. If instead Stalin had decreed that Chechnia was an independent republic like its neighbor Georgia, it too might have been spun off as a newly independent country and no one would have complained.

In addition to the political fragmentation, the breakup of the country and the disappearance of that unified economic space hit Russia very hard and pushed it toward bankruptcy. While the CIA in the 1980s once estimated that the Soviet Union's gross domestic product (GDP) was about half that of the United States, by 1992 the agency concluded that the Russian GDP had fallen to about 10 percent of the U.S. GDP. Some economists such as Simon Johnson, Daniel Kaufman, and Andrei Shleifer suggest that this is an understatement. They argue that the official statistics do not reflect the full growth of the just legitimized private sector.[1] Given the turmoil of the times, that may be true, but there is little doubt that most of the traditional industrial sectors suffered badly. By 1996, for example, petroleum production, the country's crucial sector, was off 47 percent from 1987. Some of the decline was due to poor production practices of the sort described earlier by the CIA. But even more important, the rivalry to privatize the various oil fields, refineries, and pipelines was at its peak and inevitably very disruptive. Equally discouraging, with oil prices in the mid-1990s hovering around a low $20 a barrel (in 2005 adjusted prices) there was not much incentive to increase productive capacity.

Virtually no Russian petroleum company increased production from 1990 to 1999. For many observers, it appeared that petroleum production was declining, almost as the CIA had predicted. Much of the industry was privatized in the mid-1990s and almost all the new owners seemed more interested in stripping and sending assets outside the country while they could still do so and before what many assumed would be a violent and far-reaching reaction. Capital flight from the country as a whole was thought to be on the order of $1 billion a month. To top it off, the country was racked with inflation (prices rose twenty-one-fold in 1992) and the government budget was running serious deficits because few of those who should be paying taxes did so.

The failure to pay income tax typified the problems encountered in the transition to a market-type economy. Private ownership became the new model, replacing central planning and state ownership of the country's factories, stores, and farms. In the Soviet Union, state taxes were levied as a turnover tax included as part of a product's retail price and so unnoticed by the buyer. The income tax that everyone paid also went unnoticed, having already been deducted from workers' cash envelopes before they received them. In the same way, the enterprise income tax was also automatically withheld by the state. Consequently, there was no need to file an income tax form nor send in an individual tax payment. As a result, only a few economists were aware that the Soviet Union even had taxes. That is why it was common for Russians to insist that the Soviet Union was superior to the United States not only because it had no unemployment or inflation but because it had no taxes.

PRIVATIZATION AND CHAOS

When the state transferred ownership of all those stores and factories to private owners, all that changed. Since it no longer could make deductions automatically, beginning in the late 1980s the state had to find some way to induce the new private owners as well as individual wage earners, voluntarily on their own, to send in taxes. That was something the public had never done before. Few could be expected to do so voluntarily just because some state official said they should. Given that tax rates were 30 percent or more and that the state was ill prepared to chase after tax delinquents, it was not surprising that in 2000, even after nearly a decade of private ownership, only 3 million Russians out of the 70 million who were supposed to pay taxes actually did so.[2]

Similarly, after the transition, there was as yet no market mechanism in place where producers and consumers could meet, be informed, and deal with one another. In the days of the USSR, there was no need for such a market mechanism because Gosplan, the Central Planning Agency, and the various central ministries did the job, even if poorly. But after 1991, when Gosplan and the ministries lost their power to make such allocations, Russians seeking to acquire even simple things such as a mattress, a saw, or a jacket did not know where to go. Imagine then how difficult it was for a factory director in search of a ton of coal or a specialized machine tool to find what he was looking for. Russia's retail stores had little or no experience in dealing with an independent manufacturer and supplier. Moreover, some of the supplies allocated previously by Gosplan came from factories that were now located in newly independent countries that no longer would take rubles. This was not only because it was no longer their local currency but because of the hyperinflation in Russia mentioned earlier.

For that matter, the privatization process itself was problematic. In an effort to win political support for his new bottom-up democracy, Yeltsin agreed to privatize the heretofore centrally planned, state-owned economy. The state issued every Russian citizen a 10,000-ruble voucher redeemable in newly issued shares of the enterprises being privatized. The intention was to ensure that every Russian would not only derive some benefit from the dismantling of the old system but would also have a vested interest in the success of the new market system. But after being subjected to seventy years of state propaganda against capitalism, few Russians understood why a share of stock in the new Russian companies was worth owning or had any value, especially at a time when Russia was in such a sorry economic condition. Not surprisingly, when the market value of their voucher fell to the equivalent of $25 and then $10, most Russians opted to sell their newly allocated voucher and its entitlement to a share of stock for a bottle of vodka or a few rubles. Vodka was concrete and pleasurable, rubles were tangible, but the stock was abstract and at the time little more than a piece of paper. So the vast majority of Russians had little more than a passing hangover or a few rubles to show for seventy years of communism.

Even worse, because of politics, greed, a flawed design, and corrupt implementation, a small number of investors ended up in control of most of the previously state-owned enterprises. One group of these newly rich, so-called oligarchs were former government officials. They simply took over ownership of the state properties that they had been managing as agents of the government. Another group of owners emerged from a seamier stratum of black market operators and money changers. While despised in the Soviet era for their anti-social black market activities, they nonetheless had learned how to operate in a shortage environment by mastering market practices even if they were illegal at the time. Consequently when markets and private ownership were legalized and no longer anti-social, these previously underground operators found themselves at a significant advantage. This group stood in marked contrast to the former government bureaucrats who were used to issuing decrees in the rigid world of state ownership, unconcerned by what the consumer might or might not actually want.

These former bureaucrats found themselves ill-equipped to operate in a market environment where consumers had choices and could not be dictated to.

DIVIDING UP THE SPOILS

In this chaotic environment, in a short time a growing number of these newly rich oligarchs became billionaires. But that did not necessarily mean they were good managers. Certainly none were self-made men comparable to a Bill Gates of Microsoft, Edwin Land of Polaroid, Fred Smith of Federal Express, Steve Jobs of Apple, or Richard Branson of Virgin. Even those adept at adapting to the market derived most of their wealth from seizing what had been state assets and in a large number of cases by stripping assets from those companies. These enterprises, taken over by the new oligarchs, were spun off from previously state-owned enterprises within the country's various ministries.

THE GAS INDUSTRY IS KEPT WHOLE

The case of the Ministry of the Gas Industry was different. Senior officials in the ministry fought hard to retain all the various properties within the confines of the ministry. They succeeded and in August 1989, the Ministry of the Gas Industry transformed itself intact into a corporation called Gazprom. This move kept the assets of the Ministry of the Gas Industry as a whole, ensuring that they would not be parceled out to various promoters—unlike the Ministry of Petroleum, which privatized what had been its wholly controlled producing fields, refiners, and pipelines. Initially the state owned all of Gazprom's stock but gradually sold some of its shares to private parties. Nevertheless, because the state remained the dominant share owner, the minister of the Gas Industry, Viktor Chernomyrdin, made himself president and CEO of this entity.

Chernomyrdin had served a long apprenticeship in both the energy sector and the government. He spent his early years working at the Orsk refinery, which is located not far from Orenburg in the Urals. Then after his army service, he studied at the Kuibyshev Polytechnical Institute.[3] From there he went to work for the Communist Party in Orsk and stayed until 1973 when he took a job as deputy engineer at the natural gas processing plant in Orenberg, near where he was born. He became director of that plant in 1978. His next move was to Moscow where he became an instructor for the Central Committee of the Communist Party. This trajectory put him in line to become deputy minister of the Ministry of the Gas Industry in 1982 and then minister in 1985.

This was also the year Mikhail Gorbachev became General Secretary of the Communist Party. Gorbachev began the reform process that ultimately led to the end of central planning and the state ownership of all the means of production. Anticipating the changes that were yet to come, in August 1989, Chernomyrdin transformed the Ministry of the Gas Industry into Gazprom, which became the country's first state-corporate enterprise. The state was still in control but now this control was exercised through shares of stock, 100 percent of which were initially owned by the state.

This was an early indication of what was to happen in the future. But the mass privatization program did not begin until mid-1992 after Boris Yeltsin had taken over as president. In November 1992, Yeltsin authorized the conversion of Gazprom from a wholly state-owned joint stock company into a private joint stock company whose shares could be owned by both the state and private parties. In February 1993, Gazprom began to sell its stock to the public and by 1994, 33 percent of its shares had been purchased by 747,000 members of the public, most of whom were able to obtain a Gazprom share of stock in exchange for one of the vouchers the state had issued to every Russian citizen as part of the privatization process. Fifteen percent of the stock was also purchased and allocated to Gazprom employees. For the time being, the state retained 40 percent of the shares (although this was gradually reduced to about 38 percent).

Given Chernomyrdin's success with Gazprom, in May 1992 Yeltsin chose him to be his deputy prime minister. He was promoted again in December 1992, this time to the top position as prime minister, a post he finally relinquished in March 1998, shortly before the financial crash of August 17, 1998.

When Chernomyrdin returned to a formal government position in May 1992, his deputy, Rem Vyakhirev, who had been deputy minister and then followed him to become vice chairman after Gazprom had been established, moved up again and took Chernomrydin's place as both chairman and CEO.

Like Chernomyrdin, Vyakhirev also came with considerable experience as a natural gas and petroleum specialist. He was also a graduate of the Kuibyshev Polytechnical Institute. After stints in Samara (called Kuibyshev at the time) on the Volga and Orenburg and Tiumen in West Siberia, in 1983 Vyakhirev was appointed first deputy minister of the Ministry of the Gas Industry under Deputy Minister Chernomyrdin, who would himself be promoted to minister two years later.

With Chernomyrdin as prime minister and his old deputy as CEO and chairman of Gazprom, the state did not closely regulate Gazprom. Taking advantage of this, Gazprom paid very little in the way of taxes or dividends to its principal shareholder (the state). Not only did the state see little in the way of taxes or dividends from Gazprom while Vyakhirev was in charge, but many of Gazprom's gas-producing wells, pipelines, and distribution entities were freely parceled out in unrestricted fashion to a wide collection of Gazprom executives' wives, children, and mistresses. Some of the largest spin-offs were transferred to ITERA, a company relocated from Russia to Jacksonville, Florida.

THE PETROLEUM INDUSTRY IS BROKEN UP

While the ultimate fate of the Ministry of the Petroleum Industry was very different, initially its privatization began in much the same way. The first step, in September 1991, was to transform the Ministry of Fuel and Energy into a joint stock company called Rosneftegaz (Russian Oil and Gas) (see Figure 4). But unlike Gazprom, which remained more or less whole, Rosneft was soon subdivided into what would eventually be almost a dozen more or less independent entities. Vagit Alekperov, acting minister of the Petroleum Industry, was one of the first to see the industry's potential. In November 1991, before the collapse of the USSR, Alekperov used his authority to set aside the Langepaz, Urengoi, and Kogalym petroleum fields and combine them into a package, call it LUKoil, and put himself in charge as the CEO. (Much earlier Alekperov had managed the West Siberian Kogalym region).[4]

The process of breaking out chunks of the former Ministry of Fuel and Energy continued and even accelerated after December 25, 1991, when the USSR split apart. In November 1992, Rosneftegaz was reduced to Rosneft. Two more companies, Yukos and Surgutneftegaz, were spun off in 1993. Vladimir Bogdanov took over as CEO of the latter, in essence the same producing combine he had supervised as a government manager under the Ministry. As for Rosneft, while bereft of LUKoil, Yukos, and Surgutneftegaz as of 1993, it nonetheless still produced more than 60 percent of the country's crude oil output. The raids on it were far from over, but at the time it controlled twenty-six oil-producing regional associations and twenty-three refineries.[5]

As questionable as it may have been to allow two senior ministry executives to seize ownership for themselves of the billion dollar assets they had been operating, that was almost benign compared to the way the rest of Rosneft was privatized as part of what came to be called the Loan for Shares initiative.

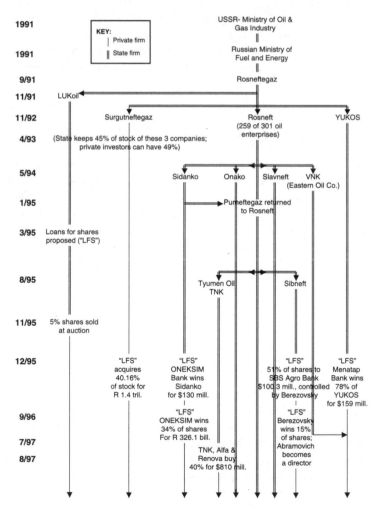

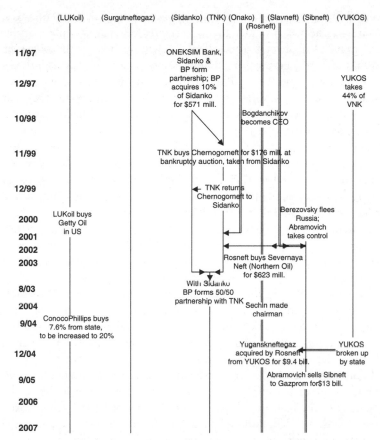

	(LUKoil)	(Surgutneftegaz)	(Sidanko)	(TNK)	(Onako)	‖ (Slavneft)	(Sibneft)	(YUKOS)
						(Rosneft)		

11/97		ONEKSIM Bank, Sidanko & BP form partnership; BP acquires 10% of Sidanko for $571 mill.	
12/97			YUKOS takes 44% of VNK
10/98		Bogdanchikov becomes CEO	
11/99		TNK buys Chernogorneft for $176 mill. at bankruptcy auction, taken from Sidanko	
12/99		← TNK returns Chernogorneft to Sidanko	
2000	LUKoil buys Getty Oil in US		Berezovsky flees Russia; Abramovich takes control
2001			
2002			
2003		Rosneft buys Severnaya Neft (Northern Oil) for $623 mill.	
8/03		With Sidanko BP forms 50/50 partnership with TNK	
2004		Sechin made chairman	
9/04	ConocoPhillips buys 7.6% from state, to be increased to 20%		
12/04		Yuganskneftegaz acquired by Rosneft from YUKOS for $9.4 bill.	YUKOS broken up by state
9/05		Abramovich sells Sibneft to Gazprom for$13 bill.	
2006			
2007			

FIGURE 4 The Breakup and Reconsolidation of the Ministry of Petroleum and Rosneft. Sources: *Kommersant* 10/23/01. Nina Poussenskova, "Rosneft as a Mirror of Russia's Evolution," *Pro et Contra Journal* 10, no. 2 (June 2, 2006); Goldman, *The Piratization of Russia*; *Russian Analytical Digest*, No. 1: Gazprom, Liberal Politics, Elections, 2006.

LOANS FOR SHARES

What turned out to be the biggest and most controversial transfer of wealth ever seen in history began in 1995 and evolved out of a proposal conceived by Vladimir Potanin. At the time Potanin was deputy prime minister under Prime Minister Chernomyrdin. Potanin proposed the Loans for Shares plan as a novel way to compensate for the fact that so few Russian individuals (especially those who came to be known as oligarchs) or businesses were paying their rightful share of taxes. Without the tax revenue, the state could not pay its bills. Under Potanin's plan, several of the banks newly opened by the oligarchs would offer to lend the government money so it could pay its bills. As collateral for those loans, Potanin proposed that the state turn over shares of stock in several of the country's petroleum companies that had not yet been fully privatized. Once the state had collected its taxes, the loans would be repaid and the collateral—that is, the shares of stock—would be returned by the bank to the state. If for some reason the loans could not be repaid, the banks, on behalf of the state, would then be authorized to auction off the collateral they were holding. After they had taken out the money they were owed, the banks would then turn the remaining proceeds over to the state.

Given the climate of the time and the rush to seize state assets, not surprisingly, this turned out to be a massive scam. Everyone knew from the beginning that there was little likelihood that the state would be able to collect the taxes it needed to repay the bank loans. How could it when the oligarchs themselves and their companies, as well as their banks, were among the largest tax delinquents? As for the auctions, almost all of them turned out to be rigged. Foreigners and most other viable bidders were excluded from the bidding. With the number of bidders sharply limited, it was no wonder that in virtually every case, the auction winner turned out to be the bank running the auction itself, or its straw or accomplice, and for a price that barely covered the amount of the loan. It was part of the Loans for Shares scheme that allowed Mikhail Khodorkovsky and his Menatep Bank to end up as the owners of Yukos that was also spun out of Rosneft, bidding a mere $309 million. (Not pocket change but cheap for even a poorly operating oil company. It soon had a market value of $15 billion.)

In a somewhat similar pattern in July 1997, Mikhail Fridman—a colorful figure whom we will turn to shortly—used his Alfa Bank and Renova, a holding company, to win control of Tyumen Oil (TNK). Subsequently, after a very contentious legal and public relations battle, TNK joined up with its one-time rival, BP, to form the TNK-BP 50–50 partnership.

Since it was Potanin's idea, it would have been unfair if he had not been able to benefit from his own program. Not surprisingly, therefore, in addition to the modest $170 million he paid to acquire Norilsk Nickel, which once privatized became one of the world's largest non-ferrous metal conglomerates (its profits in 2000 were reported to be

$1.5 billion),[6] Potanin and his OneksimBank also won control of the oil company, Sidanko, for $130 million. This had been one of the Ministry of Petroleum Industry's operating units located in West Siberia, and it, like the other privatized oil companies, was spun out of Rosneft.

The duo of Boris Berezovsky and Alexander Smolensky were more devious in their efforts. Berezovsky, who at the time had close relations with the Kremlin, particularly one of Boris Yeltsin's daughters, was behind the August 29, 1995, Presidential Edict which spun off Sibneft from the Ministry of Energy and Rosneft. Alexander Korzhakov, Yeltsin's one-time bodyguard, claims that as part of the deal, Berezovsky promised Yeltsin that if he were given ownership of Sibneft, he would then see that ORT, the TV network Berezovsky controlled, did all it could to back Yeltsin in the 1996 campaign for reelection as president. Not surprisingly, the bidding process for Sibneft was even more opaque than normal. In a December 1995 Loans for Shares auction, a heretofore unknown company FNK (Finansovaia Neftyanaia Kompaniia— Financial Oil Company), acquired 51 percent of Sibneft shares for a paltry bid of $100 million, plus the promise that more money would be invested subsequently.[7] FNK turned out to be a front for Alkion Securities, which turned out to be 100 percent owned by SBS/AGRO, which— surprise, surprise—was run by Alexander Smolensky in partnership with Berezovsky. As a further indicator of how rigged the whole process was, the auction for Sibneft was conducted by the Neftyanaia Finansovaia Kompaniia or NFK (note the similarity in name and initials), which turned out to be controlled by Berezovsky.[8]

The owners of the two already privatized petroleum companies— Vladimir Bogdanov, the CEO of Surgutneftegaz and Vagit Alekperov of LUKoil—also used Loans for Shares to enhance their personal stock holdings. But at least they had spent many years working out in the oil fields and managing petroleum production. By contrast, almost none of the future owners of the other oil companies, that is, Potanin, Fridman, Berezovsky, and Smolensky, had had much prior experience in the petroleum industry. Khodorkovsky had spent several months as a deputy minister of Fuel and Energy in 1993. But after he took over Yukos and went to look over his company's newly acquired oil fields in Nefteyugansk to "learn how the drilling process works," his host Vladimir Petukhov, the mayor of Nefteyugansk and an oilman with a doctorate in oil technology, was appalled to discover that Khodorkovsky, despite that stint in the Ministry of Fuel and Energy, had never seen an oil field before.[9]

EVERYONE WANTS TO BE A BANKER

To understand how Potantin, Berezovsky, Smolensky, Fridman, and Khodorkovsky managed to be in a position to bid for these large petroleum companies, it is necessary to detour a bit and explain how they came to establish their own banks. After all, only a few years before, none of them had any net worth to speak of.

With so little to begin with, how did they manage by 1997 to become billionaires? The explanation is that all five were able to take advantage of the Russian public's enormous hunger for consumer goods they had been denied for more than seventy years under Soviet central planning. The demand for personal computers (then a relatively new invention and in any event rare in Russia) was particularly intense. It also helped that when it became legal to establish private commercial banks for the first time in 1987, the capital requirement was the ruble equivalent of only $750,000. As trivial as this was, because of inflation by 1990 the equivalent in rubles amounted to as little as $75,000 in real terms.

The case of Mikhail Fridman is typical. The son of an academic father, Fridman, after graduating from the Moscow Institute of Steel and Alloys, worked in a steel mill for two years from 1986 to 1988.[10] Although trained to work in a Soviet state-owned factory, even as a student, Fridman began to take odd jobs on the side. Among other chores he washed windows, organized a discotheque, and did construction work. In 1987, when it became legal to set up a private or cooperative business, he opened Kuryer, a cooperative that offered such services as package delivery, window washing, and assistance with apartment rental. None of these activities required much in the way of startup capital—all they needed was labor. But as he began to accumulate a little capital, he began importing sought-after Western consumer goods, including cigarettes, perfume, computers, and even Xerox machines. He also opened up a network of photo labs. Then in a very distinct departure from such retail operations he opened a commodity trading firm. In January 1991, while Gorbachev was still president of the USSR, Fridman took his newly accumulated capital and established the first office of Alfa Bank. To do this, he needed 6 million rubles which at the time was the equivalent of $100,000, a relatively small amount for the capital of a bank.[11] It was through Alfa Bank that in July 1997 Fridman, in partnership with Access Industries—a company established in the United States by Leonard Blavatnik, a Russian émigré—was able to purchase 40 percent of Tyumen Oil Company's shares for a bid of $810 million. In doing so Fridman and Blavatnik became the effective owners of the company in much the same way Alexander Smolensky began to build his fortune by performing similar odd jobs. They too required little in the way of capital, but if such services were to be performed legally through official Soviet central planning channels, the customer would have had an enormous wait, sometimes months if not years. There seemed to be shortages of almost everything, including plumbers, carpenters, and general repairmen. For that reason, many Russians were willing to pay something extra under the table to have the work done right away. To illustrate how bothersome the shortages and delays were, the Russians delighted in telling the story about Ivan. He had been waiting and waiting for six or seven years to buy his own automobile. After waiting all that time, he finally was notified to appear July 1, 1980, at the regional office of the Ministry of Trade.

"I have good news for you," said the clerk. "Your car will be delivered to you five years from now on July 1, 1985."

"Wonderful!" Ivan replied. "But will it be in the morning or the afternoon?"

"What difference does it make?" asked the puzzled clerk. "That is five years from now."

"Well, I have to be home that morning because it's the only time I could arrange for the plumber to come."

Smolensky began to build up his fortune by specializing in construction work. The Russians had a special term for such private work crews—they were called shabashniki. While it was difficult enough to find anyone willing to do such work, it was more difficult to find work tools and construction supplies, even such simple things as two-by-four lumber and hammers and nails. Recognizing an opportunity, Smolensky began to buy up such products where he could and on occasion even manufactured these items and sold them to other moonlighting entrepreneurs.[12] All such private activities were illegal and classified as economic crimes. Eventually Smolensky was found guilty of using government printing presses to sell Bibles for private profit and sentenced to jail for two years for just such an economic crime. In 1987, when Gorbachev finally made such activities legal, Smolensky set up the Moscow No. 3 Constructive Cooperative. On February 14, 1989, two years before Fridman did the same thing, Smolensky took the rubles he had accumulated and established what he called the Stolichny (Capital) Bank. He later expanded the bank by buying up Agroprombank, which had been a state-owned bank designed to provide banking services to rural areas. Combining the two banks, he changed the name to Stolichny Bank Savings/Agro or SBS/Agro. Together with Boris Berezovsky, in December 1995 and then again in September 1996, the two men won majority control of Sibneft in one of the Loans for Shares auctions discussed earlier (see Figure 4). The SBS/Agro bid for control of Sibneft was only $100.3 million. Not bad for an asset worth upward of $10 billion.

In the case of Vladimir Potanin, he managed to build up his OneksimBank not so much by using his own labor but by subverting government agencies to his own personal ends. Like his father, Potanin worked for a Foreign Trade Organization (FTO) under the Ministry of Foreign Trade. These FTOs were set up to import and export goods on behalf of the state and to act as agents of the various state enterprises which themselves were not authorized to engage in foreign trade. Only the FTOs were allowed to have foreign currencies. In Vladimir's case, his FTO was Soiuzpromexport and it specialized in the export of nonferrous metals.[13] After he saw how others were enriching themselves with their newly created cooperatives, Potanin decided to capitalize on his own specialization by creating a cooperative that would do privately what he had been doing on behalf of the government. Leaving the government, he created a cooperative called INTERROS that began to trade in nonferrous metals. Next he decided he needed his own bank. To generate the capital he needed, he took advantage of his former government connections and supplemented his own money with borrowed funds from Vneshekonombank, the state-owned foreign trade bank. Once he had the required capital he used it to open OneksimBank, yet another in the series of personal banks established by these newly rich oligarchs.

To ensure the continuation of the privatization process, Potanin along with most of the other oligarchs used his bank to finance Yeltsin's presidential campaign effort. They all worked together to defeat Yeltsin's main rival and critic of the privatization process, Gennady Zyuganov, head of the Communist Party. As a reward for his support, in August 1996, after the election, Yeltsin appointed Potanin first deputy prime minister. After a short time, however, Potanin left office in March 1997 to go back full time into business. He did not return empty-handed. With the help of his Loans for Shares program, Potanin ended up owning twenty former state enterprises. These included not only the petroleum company Sidanko, which in late 1997 joined in a partnership with the British firm BP, but in an equally good deal, he won ownership of Norilsk Nickel, a company that produces one-fifth of the world's nickel, two-thirds of its palladium, and one-fifth of its platinum. As the profits of his company INTERROS began to grow, he expanded it into foreign markets, including the United States. There it bought up the QM Group, a nickel-producing company in Cleveland, and Stillwater Mining, a palladium and platinum producer in Montana.[14]

Given how low the price of petroleum and ferrous and nonferrous metals was in the 1990s, ownership of these Russian companies did not always look like the bargain it would become once commodity prices began to rise and oil prices hit $30 a barrel or more. Yet even in the mid-1990s, when prices were low, there was a growing awareness that the Loans for Shares scheme benefited an opportunistic and unscrupulous few at the expense of the state. More than that, the victors frequently quarreled among themselves and on occasion settled their disputes with mayhem and occasionally murder.

PLAYING HARDBALL

None of the oil oligarchs was willing to give a potential challenger the benefit of any doubt. TNK (Tyumen Oil) was particularly aggressive. In one instance that was not widely publicized, NOREX Petroleum of Canada charged that in June 2001, TNK and its parent bank, Alfa, sent in "machine gun-toting guards" to seize the production facilities of Yugraneft in Siberia. NOREX insists that at the time, it owned 60 percent of the shares of the company and that it, not TNK, was the governing partner in their joint venture.[15] TNK has justified its decision to send in its armed guards by claiming "that NOREX's capital contribution in the form of 'know how' had been improperly valued." In other words, NOREX did not own as large a share of the joint venture's capital as it claimed. Using that as a justification, TNK argued that NOREX was obligated to surrender its operating control of Yugraneft.[16] Since NOREX refused to yield control, TNK says it had no choice but to send in its armed persuaders.

In much the same spirit and in another equally brazen move, TNK also decided to take over ownership of the Chernogorneft oil fields in West Siberia. TNK wanted these fields because they were adjacent to TNK's big Samotlor Field. It made sense to combine these two fields to prevent one company from attempting to draw oil from underneath its neighbor's reservoirs. When that happens, the pressure is reduced and output in both fields is less than it otherwise would be. At the time Chernogorneft was owned by Potanin's Sidanko and Potanin's new junior partner, the U.S. oil company, Amoco, to which he had sold a 10 percent interest. Amoco in turn was purchased a short time later by BP.[17]

To gain control of Sidanco and its Chernogorneft fields, in October 1998 TNK arranged for a minor creditor to sue Chernogorneft in a provincial court for an unpaid bill of only $50,000. Two months later the local bankruptcy judge declared that because it had not paid this bill, even if a trifle, Chernogorneft was indeed bankrupt. (Once a company is declared bankrupt, it is too late for the parent company to try to pay off the debt.) The judge then assigned its assets (which were more than adequate to pay off the rather trivial $50,000) to a creditor who turned out to be a front—surprise, surprise—for TNK. The suit did nothing to enhance Russia's reputation for adherence to honest and ethical business codes, especially when it became known that the bankruptcy judge was an appointee of Leonid Roketsky, who at the time was governor of the Tyumen Region. That was not the problem. The problem was that in his off hours, Roketsky also just happened to be the chairman of TNK. Simultaneously, a straw subsidiary of TNK bought up 60 percent of Chernogorneft's debt. By the time the judge was finished, TNK had become Chernogorneft's effective owner. As a result, BP had to write off $200 million of its investment in Sidanco.

Seeking revenge, BP launched an attack on TNK that eventually involved both Madeleine Albright, then the US secretary of state, and Dick Cheney, then the CEO of Halliburton, the petroleum service company. In an effort to enhance its productivity in its oil fields, TNK signed a $198 million contract with Halliburton for the purchase of its services and access to more advanced technology. To finance this, Halliburton and TNK applied for a loan from the U.S. Export-Import Bank in Washington. Angry that an official agency of the U.S. government had agreed to underwrite what it viewed as the theft of its property, BP protested to Secretary of State Albright, who found a way to abort the loan. After some bitter recriminations on both sides, however, BP and TNK kissed and made up, and in August 2003, in the presence of Vladimir Putin and British Prime Minister Tony Blair, the heads of both companies agreed to form a TNK-BP 50/50 partnership that would operate TNK's assets under BP management within Russia. BP later encountered some problems as the Russian government adopted a more hostile attitude toward foreign involvement in the Russian energy sector, especially in cases where foreigners own as much as 50 percent of the venture. Despite occasional statements to the contrary, Putin made it very clear that while he was happy to have foreign investors put their money in Russian energy companies, he did not want foreigners running them. In September 2007, he made it explicit, complaining that too many foreigners were managing Russian companies. Reflecting that same xenophobia, there were widespread rumors that Putin would arrange for the state to buy out TNK-BP's Russian partners.[18] Then once it owned 50 percent of the TNK-BP venture, the state would move to reduce BP's share to below 50 percent or push it out completely. Dick Cheney, of course, went on to become the US vice president under George W. Bush. Before long, he found himself being blamed because the United States had become mired in Iraq, so neither TNK-BP or Cheney has lived happily ever after.

PETROLEUM OUTPUT DECLINES

Such infighting did nothing to advance the interests of the state or petroleum production. For eight years oil production continued to decline. By 1998 it was about 60 percent of what it had been at its peak. In desperation to break out of the yearly decline and in an effort to spark new output, just as it had seventy years earlier, the Russian government grudgingly allowed foreign companies such as BP to acquire an equity in Russian energy ventures, especially as it sought to develop some of the more remote offshore and Arctic locations. To make it worthwhile for Western companies to tackle the very difficult working conditions offshore near the island of Sakhalin and in northern Siberia, fields that required technology that Russians companies lacked, the Russian government agreed to sign three Production Sharing Agreements (PSA) with foreign companies, something it had been reluctant to do earlier. A PSA is more attractive to an oil company than a regular operating agreement because it allows the oil company to recoup all of its costs before it has to share any profits with the state. For the same reason, states do not like to make such concessions because they feel they should share immediately in the resulting revenue.

One PSA was offered to the French company Total as an inducement for it to undertake the development of the Kharyaga field in Timan-Pechora. According to the geologist John Grace, no Russian company seemed to be able to work with the poor quality oil in the field. The other two PSAs were offered in an effort to attract developers to the island of Sakhalin. The first PSA was signed in June 1994 with a consortium led by Royal Dutch Shell, which agreed in exchange to work the Sakhalin II offshore oil and gas fields. Because of the extreme weather, it is impossible to work there in the winter months. Shell agreed to put up 55 percent of the equity with two Japanese partners, Mitsui, which took 25 percent, and Mitsubishi, which took the remaining 20 percent. Notice that there were no Russian partners in the Sakhalin II project. By contrast, in the Sakhalin I consortium signed a year later, Sakhalinmorneftegaz was included with a 11.5 percent equity and Rosneft with 8.5 percent. But because they too lacked the technology and experience of working in Arctic offshore conditions, it was agreed that Exxon-Mobil would serve as the lead partner with a 30 percent share. The other partners were SODECO, a Japanese company with 30 percent, and an Indian company, ONGC Videsh, with the remaining 20 percent.[19]

The Russian government agreed to these PSAs with great reluctance and only because the authorities were so eager to halt the slump in petroleum and natural gas production. Russian oil companies, including Yukos and its owner, Mikhail Khodorkovsky, led the opposition to PSA concessions.[20] He and some others viewed the offer of a PSA for a foreign company as a form of unfair competition. But with petroleum prices barely rising above $10 a barrel in 1999 and output 40 percent below its 1987 peak, the prospects for a recovery in petroleum and gas production were not very good. Thus as in times past, Russia was forced to make concessions to obtain the help it needed. However, in a repeat of history, as soon as it felt confident enough to operate on its own, it moved to invalidate those same concessions.

4

Post-1998 Recovery

The Petroleum Export Bonanza

THE 1998 FINANCIAL MELTDOWN

The 1998 financial crisis hit Russia hard. There were obvious signs that the Russian fiscal system was in desperate shape, but at the time, few saw that a collapse was eminent.[1] Indeed, the almost universal conventional wisdom was that Russia had successfully managed its transition from Communist central planning to market capitalism and that the future was bright. Many had come to believe that Russia had become the next Klondike Gold Rush. They urged investors to put in their money before share prices rose even higher! Only fools and anti-Sovietchiks could think otherwise. Typical were studies such as Anders Aslund's *How Russia Became a Market Economy*, published in 1995, and Richard Layard and John Parker's *The Coming Russian Boom*, published in 1996. Both appeared just in time for investors to buy in before the financial crash that followed shortly thereafter in August 1998.[2] While the economy and its stock market have recovered significantly since then, there were many as we shall see who took their advice at the time and lost considerable sums as a result—in several cases, hundreds of millions of dollars.

It was easy to be misled. Bullish signs were everywhere. By October 6, 1997, the RTS index, the Dow Jones Index of the Russian Exchange, hit 571, an all-time high. That represented an almost fivefold increase over just a half dozen years. Investors who bought shares on October 31, 1996, in the Lexington Troika Dialog Russian Fund, which invested only in Russian companies, had a threefold increase in one year, a higher one-year return on their investment than stock market investors anywhere else in the world. Bankers in London, Frankfurt, and even New York trampled over each other to buy Russian stocks and lend money to Russian companies and government borrowers. Few could resist the frenzy.

What such analysts and investors chose to discount or ignore, however, was the deplorable state of the Russian economy. As of 1998, the officially reported GDP, as well as crude oil output, had fallen by 40 percent or more from its 1991 level. At the same time, there was also inflation. In 1992 alone, prices rose twenty-six-fold, and then more than doubled each year for a several years thereafter. By 1997, price increases had moderated to 11 percent a year, an improvement, but by most standards, still high. Overall, by December 1999, it took 1.6 million rubles to buy what 100 rubles could have purchased in December, 1990. Of course there wasn't much on the shelves to buy in 1990, but be that as it may, this hyperinflation wiped out whatever savings most Russians had built up.

Nor did it look like inflation would be less of a problem in the future. How could it be, when the government was generating an immense deficit and growing debt each year? Few Russians were paying their taxes and those that made a payment rarely paid as much as they actually owed. Combined with inflation, the underpayment of taxes meant that each year the budget deficit grew larger, which in turn meant that the government had to borrow even more money.

By mid-August 1998, government authorities concluded they could not continue what, in effect, was "kiting their checks." This involved writing a check to pay a bill from a bank account with not enough money in it at the time but with the expectation that there would be a check from another bank a few days later, which would cover the first check before it was presented for payment at the first bank. This is done in the hope that neither bank would realize that initially there had not been enough actual money to pay the bill. The Russian Central Bank and the Treasury had simply run out of money to pay the bill. When a government bond matured, the only way Russia could compensate the bondholder was to roll over the loan and issue another bond in the hope that the original bondholder would accept the new bond as a replacement for the old one and not ask for cash.

Alternatively, the government could hope that someone new would be foolish enough to buy a new government bond so the funds could be used to pay the first bondholder for the same amount he had paid for the bond, plus interest. But because the revenue the government collected was so little and slow in coming in and the interest rate it had to pay to attract lenders willing to buy those reissued government securities was so high, after a time the government was forced to borrow larger and larger sums of money just to pay the ever-increasing amount of interest. It was a marathon race without a finish line.

Since this fiscal slight of hand was unsustainable, the government eventually was forced to default on its debt. Simultaneously, it found it no longer had enough dollars to meet the demand of those who wanted to exchange rubles for dollars at the official rate of exchange. In other words, it had also run out of dollars. Unless it acquired enough new dollars, it would eventually have to devalue the ruble and require that those who wanted to buy dollars pay more rubles for them. As of mid-August 1998, because it could not find enough lenders willing to buy new or reissued government securities, combined with the fact that it had run out of dollars and convertible foreign currencies and could not pay its bills, the Russian government, in effect, had become bankrupt.

It was inevitable that the Ponzi-like scheme the Russian Treasury was running—where each day it had to find more and more new lenders so it could pay off earlier lenders—could not endure. By August 17, 1998, the Treasury and Central Bank were forced to announce that they could no longer redeem the country's bonds and pay back its lenders. This collapse was precipitated and made even more serious by the financial upheaval that hit Southeast Asia earlier in 1997. As Thailand and what had seemed to be the other dynamic economies of Southeast Asia fell into recession, commodity prices collapsed. Since most of what Russia exported was commodities, this hurt Russian export revenues. When speculators around the world sensed that Russia might also be vulnerable, they began to sell off their Russian stocks and bonds, thereby anticipating and precipitating such a collapse. This increased the hesitancy among those investors and governments who might otherwise have been willing to provide additional financial support. The IMF did provide a last-minute loan, as did Goldman Sachs, but both loans proved to be inadequate and controversial. Because Russian officials met with the owners of some of the oligarch-run banks before the government publicly announced the debt and a foreign currency exchange moratorium, some of the private Russian bankers used their insider information to sell their government securities and cash out their ruble holdings for the dollars sent in by the IMF and Goldman Sachs before outsiders could seek similar protection. This further undermined confidence in both the government and public officials.

The consequences of such domestic and international economic and financial mismanagement were far-reaching. Since government securities (that is, bonds and short-term securities called GKOs) were the main assets on the balance sheets of most of the country's banks, and since these securities were now all but worthless, most Russian banks were no longer viable. In all but a few banks, liabilities exceeded assets. Many of the oligarchs who had only recently been at the top of Russia's income pyramid found their banks were worthless. For a time it looked as if as many as 1,500 Russian banks would have to close their doors.[3] In effect, Russia found itself in the midst of a bank holiday similar to the one Franklin D. Roosevelt declared in the United States in the early 1930s. Some bankers, like Khodorkovsky, managed to survive because before his bank Menatep went bankrupt, Khodorkovsky used it to finance the purchase of properties such as the oil company Yukos, which he bought through the Loans for Shares auctions. While most of those companies were not wildly profitable, they made enough to sustain Khodorkovsky's other operations. But like Menatep, most other banks simply had to close. It did not make Menatep depositors very happy to learn that Khodorkovsky had arranged to transfer the few viable assets that remained out of his Menatep Bank into another financial entity that he operated in St. Petersburg. There they were beyond the reach of all the helpless depositors who had put their money into Menatep.

Russian industrial output and the stock market also took direct hits. The gross domestic product was 5 percent lower in 1998 than 1997. The impact on the Russian stock market was much more far-reaching. By October 1998, just a year after the October 1997 record RTS high of 571, the index fell to a mere 39. For all intents and purposes, the Russian stock market had disappeared.

The impact was not restricted only to those who had invested in the Russian stock market or to Russians. Western banks that had been lending so eagerly to Russian borrowers found themselves with worthless bonds. Some had to write off several hundred million dollars' worth of loans. Credit Suisse First Boston, for example, lost $1.3 billion and Barclay's Bank in England lost $400 million.[4] In the United States, Bankers Trust wrote off a comparable amount to that lost by Barclay's Bank.[5] More than that, there were fears that the whole U.S. financial world would be similarly affected when the monster hedge fund, Long Term Capital Management (LTCM), in Greenwich, Connecticut, acknowledged that it had lost $1.86 billion of its capital and was insolvent. It was not that the Fund itself had invested in Russia. Rather, it had lent money to other investors who were affected by the moratorium on Russian debt and the collapse of the ruble, none of whom could now repay their loans. Were it not for the timely intervention of the U.S. Federal Reserve Bank, it was likely that the LTCM collapse would have triggered a cascade of other defaults throughout the financial system. Out of concern over the impact of LTCM, the Dow Jones Index dropped by over 20 percent.

Banks and the Dow Jones Index were not the only ones affected. In the panic that followed, many foreign investors who had already set up operations in Russia—not the least of which was Pizza Hut—and imported what they sold were forced to close down. Many decided it was best simply to walk away from investments worth tens of millions of dollars. Others who were thinking of investing simply went elsewhere. Simultaneously, the price of petroleum, Russia's most important export product, fell from $26 a barrel in 1996 to almost $15 a barrel (see Table 2.1). With its banks closed, its credit worthless, and its main export product earning only 60 percent of what it had two years earlier, Russia saw many of its businesses close or come to the verge of closing, and the prospects for the Russian economy were bleak.

The drop in oil prices in the early and mid-1990s had a devastating impact on oil production. With oil prices so low, by the time the petroleum producers allowed for production costs, taxes, and transportation expenses, there was little and often nothing left over for profit. So the new owners (many of whom were now private entities) not only halted exploration for new fields, they also cut back production in existing fields.[6] As a result, Russian crude oil output fell nearly 40 percent from 1990 to 1998.

A QUICK RECOVERY

But sooner than might have been expected, the world economy began to recover. Led by an increase in commodity prices in southeastern Asia, where the recession began a year earlier, energy prices also began a quick recovery. By 2000, oil prices hit $33 a barrel, double what they had been only two years earlier (see Table 2.1). What had been a glutted market almost overnight turned into a tight market.

Much of the impetus for this change was due not only to a recovery in Europe and the United States but an ever larger increase in demand for oil and gas in India and China. Whereas China was actually a net exporter of petroleum in 1993, by 2005 it had become a major importer, forced to import 40 percent of its petroleum.[7] In 2006, it imported 138 million tons of crude oil and 24 million tons of refined petroleum.[8] Only the United States imports more. As the Chinese economy grew, this new wealth brought with it an even higher demand for petroleum. Chinese consumers' increasing use of cars and air conditioners, machines that are particularly heavy users of energy, was especially important in driving up demand. Simultaneously, China continued making massive investments in heavy industries such as steel, aluminum, and cement plants, all of which require very intense input of energy.[9] So in 2004, while China's GDP rose about 9 percent, oil consumption rose 16 percent. Overall, from 2001 to 2006 China's energy consumption rose an average of 11.4 percent annually, which was greater than the 10 percent annual growth of its GDP during similar years.[10] Oil consumption did not increase as much in the years immediately following, but still by 2006 China consumed about 7.5 million barrels of petroleum per day (350 million tons), 6–8 percent of the world's total and second only to the United States, which consumed about 20 million barrels per day (940 million tons).[11] Some Chinese economists project that energy consumption in China will more than double between 2006 and 2020 and triple by 2030.[12] This would mean China will be importing 500 million tons of petroleum a year, which approximates Saudi Arabia's entire production. (Some of this would come from countries that no longer need to consume as much because they have become more efficient in using what they have. But that would not free up enough to satisfy China's need. Who the suppliers will be for the additional coal, oil, and gas needed to feed China's voracious energy consumption is not clear.)

Recognizing their problem, the Chinese government, for example, seeks to reduce energy consumption per unit of GDP by 20 percent from 2006 to 2010. That would help, but since China grows by 10 percent a year, much of that saving would be absorbed by the higher rate of growth. Moreover, so far the Chinese have been able to reduce energy consumption per unit of GDP by only 3 percent a year.[13]

The story is much the same in India. It now imports two-thirds of the energy it consumes. The expectation is that it will have to import even more to fuel its future growth, especially if it continues to grow annually at 8 percent as it did in 2006. What makes China's and India's appetite for energy particularly important for Russia is that these newly enriched super-size populations have created an unprecedented new market situation. Their incremental demands in the early twenty-first century have sopped up most of the world's available excess oil-production capacity and more than offset whatever energy conservation may have been achieved in countries like Japan or Europe and in 2006, even the United States.[14] From 2001 to 2005, China was responsible for 30–40 percent of the increase in oil consumption. Emerging market countries as a group in 2005 generated 90 percent of the incremental growth in demand.[15] No wonder prices rose to what seemed to be new highs.

If allowance is made for inflation, 2007 oil prices were not, in fact, at record levels. April 1980 oil prices, for example, if adjusted for inflation in mid-2007 would have amounted to $101, about equal to what seemed to be the record $100-a-barrel price of January 2008. Nonetheless, in mid-2007, the International Energy Authority predicted that because of growing market pressures, real energy prices would continue to increase through 2012. As they saw it, world demand for petroleum would grow at an average of 2.2 percent a year while oil supply in non-OPEC countries would expand at only 1.1 percent. This would reduce OPEC's spare capacity and lead to continuing high energy prices.

The tightening of the market for energy products and the increase in prices that followed, even if not at a record level, had a direct and immediate impact on Russia. After a half dozen or more years of asset stripping and a corresponding reluctance to invest in new exploration and development, the oligarchs and managers of energy-producing entities came to realize that with higher energy prices they could make more money by putting their funds into exploration and production at home rather than by stripping such assets and investing the proceeds from their sale abroad. Their decision to increase production was also affected by the state's decision in 1998 to begin liberalizing taxation by instituting a flat 13 percent tax on income. It also helped that after the devaluation of the ruble in August 1998, the cheaper ruble meant that foreigners could buy more Russian products with their dollars and euros, which helped to increase Russian exports.[16]

So the oligarchs began to invest in geological exploration and better equipment. This included using more advanced Western technology. In September 2006, I had a chance to see how important Western technology has become for the Russian oil industry when I visited the Yuganskneftegaz Priobskaia oil field in West Siberia. This was the oil division that had been taken over by the state-owned Rosneft company from Mikhail Khodorkovsky's Yukos. Almost all the drilling there was being done by the American-French company, Schlumberger. Halliburton, Vice President Dick Cheney's former company, is doing much the same thing elsewhere in Russia. They both are using technology denied to the USSR during the Cold War. When the Cold War ended, the Russians still were unable to use this technology because with oil prices so low, they could not afford it. Once oil prices rose, however, Russian companies were able to hire such service companies and in doing so, they gained access to deposits that would otherwise be beyond the reach of their indigenous technology. Almost immediately there was a sharp jump in production, the first time there had been a meaningful increase since 1987. Contrary to the earlier prediction by the CIA that Russian oil production would fall off sharply, in 2000 Russian oil production rose 6 percent and by 2003, 11 percent. While the rate of growth fell to 2 percent in 2005, by 2006 Russia was even out-pumping Saudi Arabia. Just as in the periods from 1898 to 1901 and 1975 to 1992, Russia once again became the world's largest producer of petroleum (see Table 2.1).

Since Russian GDP turns out to be almost entirely dependent on changes in oil production, after years of decline Russia's GDP also increased significantly. As Table 4.1 indicates, there is an almost perfect correlation between oil production increase and decrease and changes in GDP. Moreover, with more output, there was more to export. By 2006, Russia's foreign trade surplus hit $140 billion, much of which went into Russia's currency reserves. In 2006 alone, Russia's reserves increased by more than $100 billion to a total of $300 billion by year's end. This meant that as of mid-2007, with more than $420 billion in the state treasury, Russia had the world's third largest hold-

TABLE 4.1   Russian Oil Production and GDP (% change)   ▪ Oil   ▪ GDP

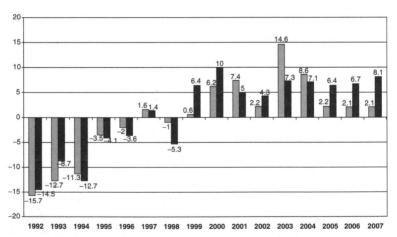

Source: "Basic Economic and Social Indicators" (for various years), Rosstat: Social and Economic Situation in Russia.

ings of foreign currency reserves and gold, behind only China, with more than $1.4 trillion, and Japan, with $900 billion.[17]

With so much cash in hand, the Russian government moved quickly to pay off its loans. As of September 2006, its foreign sovereign debt amounted to about $73 billion, less than half of the $150 billion it owed in the aftermath of the August 1998 financial collapse.[18] Much of this debt was prepaid in advance of when it was due. In August 2006, for example, Russia paid $23.7 billion to the Paris Club (creditor countries that join together to try to collect money they are owed by other debtor countries), some of it in advance of the due date.[19] Along with the buoyant yearly growth of its GDP, this prepayment helped to improve Russia's credit rating. By contrast, while the government was paying down its debt, the private corporations and banks moved in to take advantage of the more favorable credit ratings and as of October 2006 had increased their borrowings to more than $210 billion. Much of this went to corporations like Gazprom, Rosneft, and UES to finance their purchase of other properties.[20] There were fears that with private corporations seduced by so much cheap money, too much of their borrowing was being used for peripheral projects that might some day prove to be a problem. Despite the pay-down of government debt, the ratio of overall joint private and government debt to GDP increased from 19 percent at the end of 2004 to 23 percent in 2005. Nonetheless, the overall financial ratings for Russia and its corporations increased markedly from their 1998 low point.[21] In July 2006, for example, the financial rating company Fitch Ratings lifted Russia from a risky to a reasonable investment rating.[22]

Those fortunate enough to have ignored Layard and Parker's book and its advice to invest just before the 1998 financial crash but who did invest after 2000 in Russian stocks (except, of course, for Yukos) probably did quite well. By then the boom had indeed come to Russia. When Putin took over as prime minister in August 1999, the capitalized value of the country's publicly traded stocks amounted to $74 billion. By 2006, the capitalized value exceeded $1 trillion.[23]

EUROPE DIVERSIFIES

At the same time that Russia's energy sector brought prosperity to most of those who invested in it, energy imported from Russia had also become attractive to those seeking to reduce their dependence on energy from the Middle East. Given the political and military turbulence in the

Persian Gulf, Europe was eager to avail itself of a supplemental source of energy. Since Russia was part of the European continent, petroleum and gas could be delivered by an on-land pipeline as well as by ship, railroad, and highway. This meant not only a shorter journey but also one no longer vulnerable to terrorism in the Persian Gulf or Suez Canal, not to mention OPEC hijinks and 1973-type political embargoes.

The land link between producers in Russia and consumers in Europe is particularly important for natural gas customers. As we just noted, unlike petroleum, which is a liquid and thus can be delivered easily by railroad tank car, truck, and pipeline, most gas can be delivered only by pipeline. The only other alternative to pipeline-delivered natural gas is LNG, carried by expensive, specially designed ships. Railroads and tank trucks are unsuited for transporting commercial quantities of natural gas.

Given all the advantages of a natural gas pipeline, it was no wonder that despite Ronald Reagan's best efforts, the pipeline from the USSR to Europe was built. As German Chancellor Gerhard Schroeder later also noted, from an environmental point of view, natural gas would be more environmentally friendly than coal or nuclear energy. More than that, the Russian Republic had the world's largest reserves of natural gas. Initially, Germany received most of its gas from the North Sea. But since the North Sea fields were more modest in size, before long, Russia became the largest supplier of natural gas to most of Europe. As the North Sea fields, especially the gas provided by Norway, begin to decline, Russia will undoubtedly provide an ever-larger share.

Of course, there was always the danger, much as President Reagan had warned, that like some OPEC petroleum suppliers, Russia might threaten to cut off the flow of its gas for one reason or another. After all, the USSR and then Russia did just that to several of its petroleum customers. Yet except for an occasional weather-related problem, Russia has behaved honorably with most of its West European customers. This was the case even during occasional tense Cold War confrontations. More than that, when OPEC cut back petroleum production and imposed an embargo on the United States and the Netherlands in 1973, the USSR refused to participate. Instead, as petroleum prices rose, it not only continued to honor its contracts but it expanded its exports of both petroleum and gas, and by doing so, it took advantage of the high world prices resulting from OPEC's heroics. As its reputation for reliability grew, whatever hesitancy some may have had about becoming dependent on Soviet natural gas dissipated, and Soviet supplies came to be accepted throughout Europe as an integral and dependable part of the region's supply network.

WHO OWNS GAZPROM?

With the collapse of the Soviet Union and its communist system, for the first time foreign and private individuals and companies could invest and buy shares of stock in these newly privatized Russian entities, including most of those producing energy. In December 1998, for example, Ruhrgas of Germany acquired 2.5 percent of Gazprom stock for $660 million and another 1 percent in May 1999 for $210 million more. Combined with another 1.5 percent of stock it controls indirectly, Ruhrgas at one point owned or controlled over 5 percent of Gazprom stock. For a time, nonstate investors, mostly Russian entities, owned 61.63 percent of the company's stock. However, the state owned more than any other single holder and so, at least in theory, it has the right to determine management control.

Yet without 50 percent plus one share state ownership, there was always the possibility that a foreign group could accumulate enough stock to take control. When Putin became president, one of his priorities was to prevent such a possibility. Accordingly, in mid-2005, he arranged for state-run Rosneft to buy up another 10.74 percent of Gazprom shares. With these extra shares, the state or state-owned entities then held 50.002 percent of the company's shares. Another 29.482 percent was controlled by other Russian businesses and institutions. Of the remainder, 13.068 percent was held by Russian individuals, and 7.448 percent by nonresident individuals, companies, and groups.[24]

Gazprom has worked to keep tight monopoly control not only over the country's natural gas pipeline network but also over its natural gas output. Occasionally, when the Russians feel unable to master the technology required to work particularly difficult sites such as Sakhalin and the Barents Sea Shtokman field, they have agreed reluctantly to allow foreign companies to work a few such fields on their own, without Russian or Gazprom involvement. But as in the past, once their national treasury begins to overflow and new confidence builds, the Russians quickly move to circumscribe foreign involvement and invariably they take development back into their own hands.

AN OPENING FOR FOREIGN FIRMS

Because the petroleum ministry, unlike the gas ministry, was not held together during privatization as a unified whole in an entity comparable to a Gazprom, privatization provided more of an opportunity for foreign companies to set up their own petroleum-producing subsidiaries and enter into joint ventures. Philbro Energy Products created a company with the romantic name "White Nights." It was one of the first joint ventures in the post-Communist era. Its concept was a laudable one. Drilling practices in the Soviet era were notorious for their poor conservation efforts and sloppy operating methods. Against this background, White Nights proposed forming a joint venture with a Russian company to take over some of the already worked and even abandoned oil wells. They were convinced they could restore them or increase their yield by utilizing advanced Western technology. So they created a joint venture consisting of a group from Anglo-Suisse and Philbro Energy Products, a subsidy of the American company Solomon Brothers. They joined with Varyeganneftegaz Oil and Gas Production Association, whose oil wells they would be reworking. The joint venture was predicated on the assumption that without foreign assistance, output from the Varyeganneftegaz field would decline at a rate of about 25 percent a year. Anything the joint venture produced above and beyond that trend line would be considered profit for the joint venture and would be shared equally by the Russian and Western partners.

While the White Nights project succeeded in producing more than Varyeganneftegaz alone working without Western technology would have been able to do, from the point of view of the Western partners, the project was nonetheless a failure. By the time all the taxes were collected (many imposed just for the occasion), the increased transit fees deducted, and the bureaucrats properly mollified (paid off), there wasn't all that much left over to share. It was not a gratifying or unique experience.

For more than a decade, another company, Conoco, faced similar problems and similar losses.[25] Conoco entered the Russian market as early as 1989. It joined with Rosneft in a venture called Polar Lights in 1991 to develop wells in the Timan Pechora Basin not far from Arkhangelsk.[26] Conoco also formed a joint venture with Northern Territories.[27]

But after unending extortion and interference from various federal and regional government officials, particularly Vladimir Butov, governor of the Nenets Autonomous District that encompasses the Timan Pechora fields, several Western firms including Exxon, Texaco, Amoco, and Norsk Hydro of Norway abandoned similar efforts, including a joint venture called Timan Pechora Co.[28] Their run-in with Butov is a good example of the political interference that the oil companies, domestic as well as foreign, often encounter. Butov was elected governor of the energy-rich

Nenets region in the northern part of European Russia in 1996. This was despite two earlier criminal convictions. His most recent difficulty in 2002 was the result of his refusal to recognize a Moscow court order that awarded an oil field to a company other than the one he favored.

While the other companies walked away from the millions of dollars they had already invested in the region, Conoco held out. But that was largely because they were stubborn, not because they were making a profit. Among other forms of harassment, Conoco had to deal with six different local taxes, almost all of which, after a time, were increased. By 1999, they found themselves having to pay twenty different taxes.[29] The federal government also surprised them by instituting a heavy export tariff after the agreement to begin the joint venture was signed.[30] That was not all. Permission to export their output was revoked periodically. They were denied access to the export pipeline. To top it off, one of the fields Conoco had hoped to develop in the Barents Sea was suddenly transferred to a Russian firm without warning or compensation.[31] Over the decade, ConocoPhillips, as it is now called, invested $600 million in return for which it earned little and sometimes nothing.[32]

IF AT FIRST YOU DON'T SUCCEED

Despite these early difficulties, ConocoPhillips decided to try again. As company officials debated whether they should go back to Russia, ConocoPhillips, like other major energy producers, concluded that energy companies seeking new untapped reserves do not have many options, and those reserves they do find are likely to be located not only in difficult geographical areas but within politically problematic countries.[33] So after much internal discussion and debate within ConocoPhillips, in July 2004, the CEO of ConocoPhillips, James Mulva, and the CEO of LUKoil, Vagit Alekperov, met with President Putin to ask if it would be okay for ConocoPhillips to spend more than $7 billion to buy up to 20 percent of LUKoil's stock.[34] For ConocoPhillips, despite everything that had gone wrong in the past, this was worth the risk. By investing in LUKoil, they acquired access to crude oil reserves at a cost of $1.70 a barrel. As the going price at the time was around $40 a barrel, that was quite a bargain.[35]

That both CEOs thought it prudent to check with Putin in advance tells us how central Putin and his government's role have become in what elsewhere, certainly among the other non-Russian members of the G-8, would usually be a purely commercial decision. But Putin and those around him had earlier signaled considerable displeasure at what, for a time, almost seemed to be a coordinated campaign by foreign energy companies to buy up control of Russian natural resource companies. As we shall see, that was one of the reasons the Kremlin was so concerned about Mikhail Khodorkovsky and the rumors and evidence that he was trying to sell off Yukos, in whole or in part, to Exxon-Mobil and Chevron. To the nationalists in the Kremlin and to the public at large, that was a heretical if not treasonous act. It was bad enough that, also in 2003, TNK (Tyumen Oil) sold half of its interest to BP and allowed BP to become the managing partner after the merger.

The way BP became a major player in Russia makes a good case study of how hazardous such a quest in Russia can be. BP itself did not initially invest directly. Instead, in 1998, it bought up AMOCO, a U.S. company that in 1997 had bought a 10 percent share in Sidanko, a Russian oil company, for $484 million. (We discussed Sidanko's privatization in Chapter 3.) By resort to chicanery in a bankruptcy court, TNK managed to destroy Chernogorneft, a Sidanko subsidiary, which it then seized from BP/AMOCO for itself. In response, BP/AMOCO decided to play it safe and wrote off $210 million of its investment in Sidanko, not something stockholders like to hear. This was followed by vituperation and lawsuits in U.S. courts against TNK. But as is sometimes the practice in post-Soviet Russia, the fact that businesses are violent enemies one day does not preclude them from holding their noses and forming a partnership the next. Thus, in August 2003, BP and TNK agreed to reconcile their differences and, of all things, form a 50-50 partnership. This cost BP $7 billion but it made geological as well as legal sense as BP's and TNK's oil fields were adjacent to each other and coordination rather than competition would be more likely to result in the maximum volume of extraction. But as both BP and ConocoPhillips have subsequently discovered, partnerships with a Russian petroleum company are not always warm and cuddly. Because of almost unbridgeable cultural differences, not to mention the premeditated attacks on one another, as often as not the partners came to feel that their union was more like a shotgun wedding than a marriage made in heaven. Viktor Vekselberg, TNK-BP's chief operating officer and one of the main Russian owners, acknowledged as much in an interview reported in the *New York Times*.[36]

To further complicate matters, President Putin himself has criticized the BP investment. He has also referred to the Production Sharing Agreement (PSA) as "a colonial treaty" and expressed his regret that the Russian officials who authorized such arrangements had not been "put in prison."[37]

Cultural differences are not the only hazard faced by Western expatriates working for the TNK-BP partnership. The company has also had problems with Russian government authorities. Although Russia is now more open to foreign business investment—even foreign investment leading to operating and manufacturing control—than in the Soviet era, not everything has changed. The sense of paranoia and xenophobia is still very much alive. Non-Russian executives in TNK-BP, for example, are prohibited by Russian law from having access to official state data about Russia's petroleum and natural gas reserves. These reserves are regarded as a state secret; foreigners who acquire such data risk being charged with espionage. But how can anyone operate a petroleum or natural gas company without data about that company's reserves? To avoid arrest, TNK-BP buys petroleum reserve data from Western companies. John Grace reports that TNK-BP uses Degolyer-McNaughton or Miller and Lenz. Other maps are also freely available on the Internet.[38] Nonetheless, in October 2006, some Russian government officials were charged with turning over state secrets to TNK-BP employees, and some TNK-BP subsidiaries have had their state secret access licenses revoked.[39]

In yet another reflection of Russia's historic xenophobia, in October 2007 Putin complained that there were too many foreign managers in senior positions in Russian companies, especially those producing raw materials. As he put it, "a thin top management stratum dominated by foreign specialists" is the reason why Russia imports so many foreign made goods and hires so many foreign specialists.[40]

In all fairness, the way the Russian government reacts when foreign investors attempt to buy their energy resources is not that atypical of how most countries react in a similar situation. If anything, most members of OPEC, for example, are even more protective. But while Russians restrict what foreigners can do and know within Russia, they see no problem when Russian companies seek to buy energy producers in other countries. Thus, Putin helped LUKoil dedicate one of its new gasoline stations in New York City after LUKoil bought up the Getty oil filling-station network, a long established U.S. business operation. Neither the U.S. government nor the Congress did anything to hamper or limit LUKoil's acquisition. Of course, LUKoil purchased Getty's 1,300 filling stations, not its oil wells, which might have triggered a more protectionist reaction. While some Americans would likely react negatively to such foreign investments because of feelings of nationalism and fear, Russian investment in the U.S. energy sector—at least in petroleum production, refining, and servicing—is a good idea. The Russians are more likely to export petroleum to the United States and avoid any halt in deliveries if they have operations within the country. Otherwise, some strategists argue, in the event of an embargo these facilities would have to be closed down. At the same time, of course, the properties Russians buy in the United States can serve as hostages if that should ever be necessary to offset similar pressure on U.S. companies in Russia. In any event, U.S. investment in Russian companies and Russian foreign direct investment in the United States symbolizes Russia's emergence as an economic and a political player of consequence.

IS THIS JUST ANOTHER BLIP?

Given how often runups in the price of energy have been followed by rundowns, might the high energy prices of 2006–2007 be just a temporary increase? During almost all of the previous energy price hikes, it certainly seemed that higher energy prices were here to stay. For that matter, when energy prices subsequently fell, few thought prices would rise again. As a glance at Table 2.1 suggests, however, price cycles, with their ups and downs, appear to be an inescapable part of the world's economic energy life.

The way economics works, it is to be expected that almost all economists would insist that energy price cycles are inescapable. Energy markets can be likened to the corn-hog cycle that economists teach to their students. When corn is scarce corn prices rise, which makes it too expensive for many farmers to breed hogs. So they kill their hogs, which reduces the demand for corn. This precipitates a drop in corn prices, which makes it cheaper to breed hogs so the demand and the price of corn rise. In much the same way, although you can not grow or kill oil wells like you can breed or slaughter hogs, when energy prices rise, energy becomes too expensive for some users who then look for substitutes or cut back. Not only are there fewer buyers (less demand) at higher prices, but the higher prices stimulate suppliers to offer more for sale. They want to sell more not only to earn a higher profit but also because the higher prices make it profitable to develop substitutes or to open up marginal sources of supply where heretofore the costs were too high to operate profitably.

While the supply and demand process needs no human organizer or controller to make it work, the Saudis have traditionally sought to ensure that swings in the price of petroleum were not too extreme. Consequently, when crude oil prices fall too low, they lobby the other members of OPEC to reduce output in order to tighten supplies and nudge up prices. On occasion, they have acted unilaterally. Similarly, when prices climb too high, the Saudis use their standby capacity to increase output because too great a price rise would stimulate the search for substitute fuels and conservation, measures that could prove hard to undo.

The rapid growth in prices in the early 2000s induced just such a Saudi reaction. After skyrocketing from $15 a barrel in 1998, to $77 a barrel in July, 2006, oil prices leveled off and for a time in early 2007 fell to slightly under $50. At that point the Saudis responded by curbing production by almost 1 million barrels a day (50 million tons) to prevent a further slide in prices.[41] But since the Soviet Ministry of Petroleum and now the Russian oil companies are not part of OPEC, Soviet and then Russian producers have traditionally tried to take advantage of Saudi and OPEC cutbacks by doing just the opposite. When OPEC has reduced output the Russians usually have increased theirs. That explains why in late 2006 when the Saudis reduced their output, Russia once again became the world's largest producer of petroleum.

In post-1998, however, there was something different about the way energy producers and consumers reacted. Producers did increase and reduce output in tandem with price increases and decreases (at least OPEC producers did), and the higher prices did revive interest in and production of renewable biofuel energy substitutes such as ethanol. Yet as prices approached the $100 a barrel mark, there seemed to be a new factor pushing prices to that level. There seemed to be less and less slack in the market. According to calculations of Fatih Birol, the chief economist at the International Energy Agency, the world needs 5 million barrels a day (250 million tons) of spare oil production capacity to avoid energy disruption.[42] That is equivalent to almost half of Russia's annual production. In 2005, there was only 1.5 million barrels (75 million tons) spare capacity.[43] Paolo Scaroni, CEO of the Italian energy company Eni, estimates that as of 2006 the world's spare petroleum capacity had fallen from 15 percent of world consumption to 2–3 percent.[44] That suggests that energy prices are unlikely to drop in the near future. What remains to be seen is what sources of supply that before were too marginal will now become profitable and how extensive such new projects will prove to be.

THE NEW DEMAND EQUATION

Equally important, not only did there seem to be less spare capacity but energy consumption seemed to be increasing faster than normal. According to an estimate by Edward Morse, chief energy economist at Lehman Brothers, the investment banking firm, overall world energy demand rose by 10 million barrels a day (500 million tons) from 2000 to 2006.[45] Subsequently, the high petroleum price in 2006 precipitated a drop in consumption of 100,000 barrels a day within countries that belong to the Organization for Economic Cooperation and Development (OECD), especially the United States. But as we noted earlier, demand in the developing countries, especially in China and India, rose faster than elsewhere thus offsetting that drop. While petroleum consumption in the United States has risen by 17 percent since 1995 to a massive 20.7 million barrels a day, the comparable figures in India were a 57 percent increase to 2.5 million barrels a day and for China, now the world's second largest consumer of petroleum, there was a 106 percent increase to 7 million barrels a day.[46]

It is always risky to predict the future, especially when it comes to the discovery of new energy supplies and energy prices, but the recent very rapid growth in demand within the developing world is unlikely to abate. As the GDP continues to rise rapidly in countries like India and China, their energy consumption is likely to grow even faster as their new wealth brings an even faster demand for energy intensive products such as automobiles, refrigerators, and air conditioners. Because all three items are considered to be icons of the middle class, demand for such products is especially strong. Given that each Chinese consumes the equivalent of two barrels of oil a year and that each American consumes twenty-six barrels, the odds are that even with higher prices, China will substantially increase its energy per capita consumption; this means that future worldwide energy demand will continue to increase rapidly and outpace discovery of new energy supplies.[47] It is this demand and supply dynamic that enhances the financial and political clout of energy-rich Russia. Undoubtedly, as is in the past, sometimes there will be an increase in supply and a slowing of demand growth (and even occasionally a decline), but it is the coming of affluence to India and China that changes the equation. As their demand for energy continues to grow, this will provide enormous economic and political opportunity for Russia.

ARE RUSSIAN RESERVES LARGE ENOUGH?

With this new dynamic, future energy markets and supplies are bound to be tighter and substitutes and supplemental supplies harder to find. While this strengthens the hand of all energy producers (and partly explains the behavior and danger of someone like Hugo Chavez in Venezuela), it is particularly important for Russia. Russia is doubly blessed. While its proven reserves of 10.9 billion tons of petroleum or

79 billion barrels (6 percent of the world's reserves) are not nearly as large as Saudi Arabia's 36 billion tons (264 billion barrels), they nonetheless make up 42 percent of the non-OPEC country reserves. Moreover, much of Russia remains unexplored by geologists, and while it is unlikely that there are any giant fields left to be discovered, given high enough prices and the right time and infrastructure there is probably still more petroleum to be found.[48]

In more recent times, as the country has allowed in Western petroleum companies and their more advanced technologies, companies like BP have found that the reserves they purchased were actually larger than they and the previous Russian operators had originally thought.[49] In April 2004, Lord John Browne, then CEO of BP, indicated that TNK-BP, which officially reported it had proven oil reserves of six billion barrels, could actually have considerably more. Although most geologists think it unlikely, Lord Browne said the total could be as high as 30 billion barrels. Robert Dudley, CEO of TNK-BP, predicted that the enhanced recovery techniques used in the West alone would make it possible to increase output by 750 million barrels. Most of the higher estimates result from advanced technology: when BP, with its Western knowledge and equipment, was able to put to work its "stronger pumps and more powerful tools," it was able "to crack open the underground sandstone," which holds in the crude oil and which TNK on its own could not tap.[50] The expectation is that as technology continues to advance, there will be similar happy surprises.[51]

Even if no large reserves are found, the present reserves are enough to provide Russia with an enormous financial windfall. As a look at Table 4.2 indicates, each year Russia generates an enormous trade surplus. In 2006, for example, the surplus amounted to $140 billion. That contrasts with $20 billion in 1995, when petroleum prices were much lower. Petroleum exports were $140 billion in 2006, which accounted for almost half of the overall export earnings and the entire trade surplus. Strategically, petroleum has brought Russia unaccustomed wealth. In addition to over $120 billion in its Stabilization Fund in 2007, it also held over $420 billion in the treasury and Russian Central Bank, which as we saw earlier in this chapter provides Russia with the world's third largest stash of dollars, gold, and convertible currencies.[52] This cash windfall has allowed it to prepay its debt to its creditors in the G-8 countries and several other groups. Not bad, considering that in 1998, a bare nine years earlier, the vault was effectively empty.

While its petroleum exports provide Russia with its new financial wherewithal, it is natural gas that brings Russia unprecedented political

TABLE 4.2   Russian Exports and Imports ($US, Bill.)

	Exports	Imports
2007	316.5	198.1
2006	303.926	164.692
2005	245.255	125.123
2004	183.185	94.834
2003	135.403	75.418
2002	107.247	60.966
2001	103.192	53.764
2000	105.5	44.9
1999	74.7	40.4
1998	73.9	59.5
1997	86.9	72
1996	89.7	68.1
1995	82.4	62.6
1994	67.4	50.5

Source: "Basic Economic and Social Indicators" (for various years), Rosstat: Social and Economic Situation in Russia. Accessed via ISI Emerging Markets, www.securities.com.

clout. Combined, the need for these two commodities makes Europe very dependent on Russia. At the same time, some Europeans insist that the Russians are equally vulnerable. As they see it, once an expensive pipeline is built and natural gas deposits developed, Russia will be as dependent on its customers in Western Europe to buy and pay for that gas as Europe will be to have access to it.[53] That may be true, but only as long as Europe acts as a united bargainer and no European country seeks to sign a private agreement with Russia. It also assumes that Russia cannot find an alternative customer in need of natural gas. That also assumes that neither Gazprom nor Russia has a leader capable of trapping the Europeans into playing off against one another or of finding other large customers in a world frantic to assure themselves of secure energy supplies. As we shall see shortly, that mind-set seriously underestimates the analytical insights and talents of those in the Kremlin, especially Vladimir Putin or his successor. It also seems to overlook the Russian penchant for chess and the ability to check the moves of their opponents. Just how premeditated and masterful Putin has proven to be will be the subject of Chapter 5.

5

Putin Takes Over
The Return of the Czar

IT WAS NOT THE BEST OF TIMES

It was not an auspicious beginning for the new prime minister. Having been the head of the FSB (the modern day KGB), Vladimir Putin certainly was aware of the problems confronting his country, but awareness of problems is not the same thing as coping with them. And Russia had problems. In August 1999, it was still reeling from the financial bloodletting of August 17, 1998, twelve months earlier. In the wake of the government default on its debt, most of the country's larger private commercial banks had shut their doors—some, such as Togobank, forever. Millions of Russians lost their savings, including former president Mikhail Gorbachev and the director of the Marinsky Opera of St. Petersburg, Valery Gergiev.

For the Marinsky, this was nearly catastrophic. Gergiev had set aside $2 million in his bank to pay for the ensemble's trip to China. Now the bank was closed and its money gone. Had Phillips Electronics of the Netherlands not come to the rescue with $1 million, the famed St. Petersburg ensemble would have been forced to cancel its tour. Others without such a fairy godcorporation to turn to were not so fortunate. Businesses closed down, staffs were fired, and the whole concept of a market economy was cursed.

Investors fared no better. They watched helplessly as their portfolios went to zero. The RTS index (the Russian counterpart of the Dow Jones Index of Russian stocks) had been one of the world's best performing indices prior to October 1997. But that was then. As we saw, from a high of 571, the index fell to 39 by October 1998, a mere twelve months later.

It was not only the well-off who suffered. As industrial output declined and unemployment increased, the number of Russians below the poverty level, which had fallen to 21 percent in 1997, suddenly soared to 33.3 percent, a new high. Western companies exporting to Russia were also hit. With the devalued ruble, few Russians could afford the cost of imported dollar- or euro-denominated products.

If it was not a good time to be in business, neither was it a good time to be in government. Within the subsequent twelve-month period, President Boris Yeltsin went through four prime ministers. Looking for a scapegoat, a week after the crash on August 23, 1998, Yeltsin fired Sergei Kiriyenko, a financial specialist and the presiding prime minister at the time. He was replaced with Evgeny Primakov, who lasted eight months—until May 1999. Yeltsin then appointed Sergei Stepashin. After barely three months in office, Stepashin was also pushed aside. His replacement was Vladimir Putin. Unlike Primakov and Stepashin, both of whom had also headed the FSB, as the KGB subsequently became known, Putin apparently was more amenable to ensuring Yeltsin that he and his ambitious daughters would be guaranteed legal immunity from any future investigation into contract kickbacks. From all reports, they needed such protection. It was widely rumored that the daughters had pocketed tens of millions of dollars from Swiss companies that had won contracts to refurbish the Kremlin and the Russian White House, among other projects. Reflecting the temper of the times, jokesters enjoyed recounting what happened to the Moscovite who one day drove his car into the Kremlin compound and parked it. Immediately a policeman ran up to him shouting, "You can't park your car there. That's right underneath Yeltsin's window!" "Don't worry, don't worry," calmly replied the driver. "I've locked the car."

One month after his appointment as prime minister, Putin moved immediately to tighten control. He ordered government troops to return to Chechnia to reassert Russia's authority there. This was done in response to the bombing of some Russian apartment houses by what appeared to be Chechen terrorists as well as the incursion into the adjoining province of Dagestan by a Chechen group led by the Chechen leader Shamil Basayev. Who actually bombed the apartment building remains in dispute. Some, such as the one-time oligarch Boris Berezovsky now in exile in London, insist that the available evidence implicates the FSB, not the

Chechens. Whoever the actual bombers were, Putin used the apartment bombing, as well as the Dagestan invasion, to justify stronger measures from the Kremlin. By doing so, he put an end to dreams of any other secessionist malcontents in the regions who might have entertained similar notions of establishing an independent country.

THE ECONOMY RECOVERS

Putin's determination to reestablish Moscow's dominance over some of Russia's restless regions was enhanced by the fact that five months before his appointment the economy began to improve. As Table 5.1 shows, in September 1998, industrial production was 15 percent below production of September 1997, but by March 1999 it once again began to grow.

TABLE 5.1   Monthly Changes to Industrial Production

Month	As % of Corresponding Period in Previous Year
January 1998	102.9%
February 1998	101.2%
March 1998	102.5%
April 1998	101.1%
May 1998	97.2%
June 1998	97.1%
July 1998	91.0%
August 1998	88.4%
September 1998	85.0%
October 1998	88.3%
November 1998	90.6%
December 1998	93.3%
January 1999	97.6%
February 1999	97.0%
March 1999	100.4%
April 1999	100.6%
May 1999	106.0%
June 1999	109.0%
July 1999	112.8%

Source: "Basic Economic and Social Indicators (for various years), Rosstat: Social and Economic Situation in Russia.

By August 1999, when Yeltsin made Putin prime minister, industrial production was already roaring along. In May 1999, for example, industrial output exceeded that of the previous May by 6 percent. Fortunately for Putin, he took office just as the Russian economy began to benefit from the recovery in Southeast Asia, the region that had triggered the economic downturn the year before. In all fairness, if anyone deserves the credit for the economic upturn, it should be Primakov because the improvement began in early spring 1998 when he was prime minister. But more than the actions of any one prime minister or Kremlin official, the best explanation for the recovery is that the recovery in commodity prices, particularly petroleum prices, made the difference.

Because of the increase in oil prices, Russia's rebounding economy would make whoever was in office at the time look like an economic genius. To his credit, Putin did nothing to hamper economic growth. On the contrary, he brought in some of his talented associates from St. Petersburg, such as German Gref and Alexei Kudrin, and put them in charge of reviving the economy. They had previously worked with Putin on economic and financial matters in the St. Petersburg governor's office and were regarded as competent technocrats. (We can include them as "FOP," Friends of Putin.) Following their advice, Putin introduced a flat 13 percent income tax and proposed a series of other initiatives, including a program to simplify and reduce the bureaucratic maze that entrepreneurs had to fight through before they could open a new business.

While the benefit of a flat tax as a stimulus to economic growth is hotly debated in the United States, it appears to have done little damage when it was introduced in Russia; to the contrary, based on the way Russian GDP grew, it seems to have had a positive impact. Previous to its passage, the maximum tax rate was set at 30 percent. Few Russians were paying any tax, much less their required share. Clearly, a low tax was better than no tax. With the rate at only 13 percent, Russians had less incentive to cheat.

In a departure from the propaganda of the Soviet era, Putin also insisted on acknowledging the seriousness of the country's economic condition. While Russia may have thought of itself as an economic superpower in the Soviet era, by 2000, Russia's per capita income was actually lower than Portugal's, then the poorest member of the European Union. Putin acknowledged that if Russia were ever to catch up, it would have to double its GDP in ten years. It could do this, however, only if it increased its GDP by at least 7 percent a year, a goal that he set for the country. Except in 2001 and 2002, Russia did come close to this although its growth was consistently due to high world energy prices more than a revitalized manufacturing sector (see Table 4.1).

The improvement in Russian GDP certainly added to Putin's popularity. Yet the GDP is still not large enough to provide for a superpower's military force, at least not one that would measure up to previous Soviet standards. Nonetheless, Putin has substantially increased the size of Russia's military budget, by 27 percent in 2005 and 22 percent in 2006. But unless he severely curbs consumption, Russia will not be able to afford the large funds needed to support its superpower military ambitions.

A ROAD MAP TO SUPERPOWER STATUS

Putin's concern for Russia's struggling economic and lost superpower status long predates his appointment as prime minister. In a dissertation submitted in June 1997 to the St. Petersburg Mining Institute and in a subsequent article "Mineral'no-syr'evye resursy v strategi razvitiia Rossiiskoi ekonomiki," published in *Zapiski Gornogo Instituta* in 1999 and translated by Harley Balzer in *Problems of Post-Communism* in January 2006, Putin outlined a plan, a sort of "owner's manual" for Russia's recovery and return to economic and political influence. The thesis itself was probably written just before and after his boss Anatoly Sobchak, governor of St. Petersburg, lost his reelection in 1996. Since Putin worked for Sobchak, this loss meant that Putin was also without a job.

In his dissertation Putin called on the Russian government to reassert its control over the country's abundant natural resources and raw materials. "The process of restructuring the national economy must have the goal of creating the most effective and competitive companies on both the domestic and world markets." He viewed this as probably the best way to reestablish Russia's status as a superpower, an energy superpower. Instead of allowing the country's oligarch-controlled corporations to focus exclusively on making a profit, Putin proposed that they should be used instead to advance the country's national interests. To reclaim some of the assets spun off to private interests under Yeltsin, Russia should commandeer these companies and once again integrate them vertically into industrial conglomerates so they could compete better with Western multinational corporations such as Exxon-Mobil and Shell. In Putin's words, "Regardless of who is the legal owner of the country's natural resources and in particular the mineral resources, the state has the right to regulate the process of their development and use. The state should act in the interests of society as a whole and of individual property owners, when their interests come into conflict with each other and when they need the help of state organs of power to reach compromises when their interests conflict."[1]

In Putin's thesis, he acknowledged that Russia would have a hard time becoming a competitive manufacturer. In a subsequent article, he also warned that if Russia's economy continued to be isolated too long from world markets, its technology would never be competitive.[2] Even in 1997, when the economy seemed to be booming, it needed large injections of capital to help develop those resources. To attract that capital, he proposed that Russia open its heretofore closed doors to foreign direct investment. Russia should welcome the infusion of foreign capital investments, but those investors must understand that Russia would retain operating control, investment or no investment. He stressed, however, that no matter who legally came to own Russia's commodity-producing companies, whether private parties or foreign corporations, the state should coordinate and regulate their activities. As he saw it, if left on their own, private owners become too absorbed in pursuing their own interests and are more interested in damaging their competitors than helping the state. They become so self-centered they ignore legitimate state interests. He insisted that it is a mistake to rely on the private owners and markets alone.[3] When Russia did that in 1991, the country's production suffered badly. Moreover, private monopolists obstruct innovation.[4] By redeclaring control if not ownership, particularly of these resource-based companies, Russia, he argued, has the potential to emerge "from its deep crisis" and restore "its former might."[5]

NATIONAL CHAMPIONS

This thesis was written considerably before Putin became head of the FSB. No one, including Putin, could have dreamed in 1996 or 1997 that he might someday be appointed prime minister as he was in August 1999, much less acting president five months later. In this thesis, Putin emphasized the concept of what he and others have come to call "national champions." But Clifford Gaddy of the Brookings Institution in the United States, has found that this notion of "national champions," which became so important during Putin's presidency, actually did not originate with Putin. In a remarkable piece of textual detective work, Gaddy and his Russian assistant, Igor Danchenko, discovered that almost sixteen pages of Putin's dissertation, "The Strategic Planning of the Reproduction of the Resource Bases," were copied almost intact from an earlier 1978 study entitled, "Strategic Planning and Policy," written by two University of Pittsburgh analysts, William King and David Cleland.[6] Their book was subsequently translated into Russian and Putin includes it in his bibliography, but there is only a single citation of it in the text.[7]

Regardless of whose idea it was originally (Charles de Gaulle advocated something similar when he was president of France in the 1950s), as soon as he became president, Putin took it as his own and began to create his own national champions. As he envisaged it, these national champions would put promotion of the state's interest over profit maximization. At home that might mean keeping energy prices low as a form of subsidy for the public. Outside Russia, it might mean suspending deliveries to countries that refuse to support Russian foreign policy or advance its interests. These national champions would most likely be more than 50 percent owned by the Russian government. But with the right type of guidance and pressure, there was no reason that predominantly private companies could not also serve as national champions. Should there be times when a private company might decide to rebuff state guidance, the state should use its powers to induce compliance. That might involve sending in state tax auditors or inspectors from the environmental agencies to check for wrongdoing. In the case of petroleum or gas producers, refusal to go along with the state or advocating undesired initiatives could be remedied by refusing such mavericks access to Russia's oil and gas pipeline monopolies that control shipment to both domestic and foreign markets.

FROM BLUEPRINT TO ACTION

As a first task in initiating his national champion program, Putin staffed Russian state-owned companies with leaders who would be more amenable to doing his and the state's bidding. This meant that he had to remove some of Russia's more notable and powerful oligarchs from their only recently privatized companies. As an indicator of Putin's success in reclaiming the state's ownership of the country's oil output, when he took over as president in 2000, the state's share of total crude oil production was 16 percent; by late 2007, it had increased to about 50 percent.[8]

Almost immediately after his election as president in March 2000 Putin set to work. Just three months later in June 2000, he forced Viktor Chernomyrdin out of his sinecure as chairman of Gazprom's board of directors, a post he had acquired only a year earlier in mid-1999 (see Table 5.2). In the Soviet era, Chernomyrdin had been minister of the Gas Industry. In 1989, only two years before the collapse of the Soviet system, he took the initiative in transforming his ministry into "Gazprom Konsern," making himself its president in the process. In late 1992, Gazprom Konsern was carried one step further and became the Russian joint stock company, Gazprom (RAO Gazprom).[9] Described by Jonathan P. Stern as the "partly privatized joint stock company," RAO Gazprom in February 1993 was in turn transformed into OAO Gazprom, an Open Joint Stock Company.[10] As we saw in

TABLE 5.2   Vladimir Putin Elected President March 2000; Quickly Begins Purges to Create National Champions

	Former Nomenclatura	Upstart Oligarchs
Viktor Chernomyrdin	6/2000 Removed as chairman of Gazprom	
Vladimir Gusinsky		6/2000 Jailed and removed as head of Media-Most
Boris Berezovsky		11/2000 Threatened with jail; yields Sibneft and flees to England
Rem Vyakhirev	5/2001 Removed as CEO of Gazprom	
Viktor Gerashchenko	3/2002 Removed as chairman of Russian Central Bank	
Mikhail Khodorkovsky		10/2003 Jailed and Yukos seized by state

Chapter 4, until mid-2005 when Putin arranged for the state to buy 50 percent plus one share of Gazprom's stock, the Russian government held only 35–40 percent of the company's shares.[11]

Appointed by Yeltsin as deputy prime minister of Russia in mid-1992, Chernomyrdin made sure that in his absence Gazprom was well provided for. Rem Vyakhirev, who had served under Chernomyrdin as vice chairman of the Ministry of the Gas Industry, succeeded Chernomyrdin at Gazprom and in mid-1992 became its CEO. A few months later Yeltsin promoted Chernomyrdin to the post of Russia's prime minister. Given the incestuousness of all these arrangements, it was not much of a surprise to learn that Gazprom under Vyakhirev became one of the largest financial angels backing Chernomyrdin's party in the December 1995 election for the Duma. It did the same a few months later in June 1996 when Yeltsin ran for reelection as president. Reflecting the closeness of their relationship, wags transformed Chernomyrdin's party slogan, "Nash Dom, Vash Dom (Our Home, Your Home)," into "Nash Dom, Gazprom (Our Home, Gazprom)."

By March 1998, Yeltsin had begun to suspect that Chernomyrdin was taking his job for granted and on a growing number of occasions had begun to act as if he were president, not Yeltsin. Consequently, Yeltsin removed him as prime minister. To soften the blow, Yeltsin made Chernomyrdin chairman of Gazprom, a homecoming of sorts. That was all pre-Putin.

The care and feeding of the Gazprom executives that characterized the Yeltsin and Chernomyrdin years changed abruptly in June 2000 after Putin won election as president. At the time it did not seem as though Putin was accomplishing very much, but looking back at his first year in office, the firing of Chernomyrdin that June was just the kickoff of a concerted campaign.

During the same month, Putin also went after the first of the "upstart" or non-"nomenclatura" oligarchs (see Table 5.2). For the most part, these were newly rich oligarchs who in Soviet days had never been included in the party or government hierarchy, officially referred to as the nomenclatura. In fact, most had one non-Russian parent and in some cases should not have been listed as Russian on their internal passports, an important prerequisite for anyone in Russia seeking inclusion on the nomenclatura list that identified who was important in the Soviet Union. And as we saw, many had also been involved in private or black market activities—what the Soviet Union classified at the time as "economic crimes."

RECLAIM THE TV NETWORKS

One of Putin's first targets among this group was Vladimir Gusinsky. He headed Media-Most, a media company that encompassed NTV, the country's largest private TV network, as well as several newspapers and magazines. Gusinsky had created that media empire in just a few years. To a Kremlin unused to a TV network that was not controlled by the state, Gusinsky had a well-deserved reputation in the Kremlin as "a pain in the neck." Yeltsin was the first to feel his bite. Gusinsky's NTV was particularly critical of the Yeltsin government's 1994 war with the Chechens. In retaliation, in December 1994 Yeltsin's bodyguards, led by Alexander Korzhakov, physically attacked Gusinsky's bodyguards, forcing them to lie in the snow outside Gusinsky's office in what became known as "the faces in the snow" incident. But as angry as Yeltsin and his family were over the way they were criticized and satirized by NTV, particularly in *Kukely*, a weekly puppet television show, Yeltsin never moved to close Gusinsky's company or jail him.

As Gusinsky would soon learn, Putin had a much thinner skin. After Putin sent troops back into Chechnia in 1999, NTV resumed its criticism—only this time Putin, not Yeltsin, was the target. Gusinsky understood there was a change in leadership and understood there was a risk to him and his media empire. In a conversation in Moscow in March 2000, the week before Putin was elected president, Gusinsky acknowledged to a group of us from the Jamestown Foundation that such outspoken criticism of Putin might cause problems for him, his staff, and his network. But he and one of his senior assistants insisted at a late-night dinner before Putin's expected victory that they would not pull their punches.

Perhaps they should have. Three months later in June 2000, only just installed as president, Putin had Gusinsky arrested on charges of embezzling funds from a St. Petersburg company.

STAY OUT OF POLITICS

In contrast to Yeltsin's tolerance of criticism, Putin summoned twenty-one of the country's new oligarchs to a Kremlin meeting convened the next month on July 28, 2000. Neither Gusinsky nor Berezovsky was invited. Had they been there, they would have heard Putin tell those in attendance that if they kept out of politics, he would leave them alone and not question how they had managed to accumulate so much wealth so quickly. His message was an implicit warning to avoid Gusinsky-type attacks in the media and interference with Putin's policies in the Duma. Most of the oligarchs got the message and paid heed to Putin's warnings.

Two oligarchs, Boris Berezovsky and Mikhail Khodorkovsky, did not. Even though he was not there in person, Berezovsky quickly learned of Putin's warnings, but typical of the arrogance of the oligarchs who rose to affluence in the Yeltsin years, Berezovsky acted as if it made no difference. When Russia's nuclear submarine Kursk sank in August the following month, ORT, the TV network Berezovsky controlled, joined with Gusinsky's NTV (Gusinsky had been released from jail) to criticize the accident and the government's belated response to it. ORT made a special point of interviewing the bereaved families of the dead sailors in their drab quarters in Vidyayevo, the submarine's home port on the Barents Sea. Where, the families wanted to know, were Putin and other senior government officials? Why weren't they at the scene of the accident? Both ORT and NTV provided the answers with video shots of Putin enjoying himself on vacation outside his home along the ritzier Black Sea. That did it. Now Berezovsky and Gusinsky were in serious trouble. Soon after, Media-Most, Gusinsky's holding company, was seized from him (ostensibly for his failure to repay a loan). He then fled to Spain and into exile in the United States and Israel.

Berezovsky can be forgiven for thinking that he would not become a Putin target. As one of Putin's original backers for the post of prime minister, Berezovsky evidently assumed that as a minimum, out of gratitude, Putin would not turn on him. After all, before all this trouble began, Berezovsky had even gone so far as to welcome Putin and his family as houseguests in Berezovsky's mansion on the French Riviera.[12] Moreover, Berezovsky had been close to Yeltsin's family and other senior officials in the Kremlin. He had become a financial supporter and confidante to Yeltsin's daughters and their husbands. Berezovsky made one of them, Valery Okulov, the CEO of Aeroflot. That in large part explains why they, in turn, agreed to set aside Sibneft, a petroleum complex in one of the Loan for Shares auctions so that Berezovsky would emerge as the dominant owner. Berezovsky, in turn, agreed to use some of the revenue from Sibneft in an off-the-books pass-through to underwrite Kremlin expenditures and Yeltsin's 1996 campaign for reelection.

Berezovsky's biggest mistake, however, was that he allowed ORT to join with NTV in its various attacks on Putin. Eventually that set off rumors that Putin had put out an order for Berezovsky's arrest. Taking the hint, Berezovsky fled to London and surrendered control of his media assets as well as his petroleum company, Sibneft, to what had been his junior partner, Roman Abramovich. Abramovich, in turn, was happy to be cooperative and graciously put them at Putin's disposal. In a few months, state-owned Gazprom took possession of Gusinsky's company Media Most and Berezovsky's Sibneft, in effect nationalizing them both. Putin's national champions were quickly beginning to take shape.

RECLAIM GAZPROM

The next year in May 2001, Putin continued his campaign by asserting firmer control over state-controlled Gazprom. He did this by using the Gazprom stock owned by the state to vote to oust Vyakhirev as CEO. With the removal of both Chernomyrdin and Vyakhirev and their replacement with Dmitri Medvedev and Alexei Miller, two younger bureaucrats who had worked with Putin in St. Petersburg, Putin was now in a position to halt the blatant asset stripping that had characterized Chernomyrdin and Vyakhirev's almost decade-long raid on Gazprom, the company they were supposedly leading.

One of the most brazen examples of this asset stripping was the way Gazprom executives aided and abetted the formation of ITERA. This company soon became the second largest producer of natural gas in Russia. ITERA stands out because although its main business was in Russia, it was headquartered in Jacksonville, Florida. As far as is known, Jacksonville was picked because it is warmer than Moscow and because the CEO, Igor Makarov, had a Russian friend who emigrated to Florida and suggested that Makarov open an office nearby. In retrospect it was probably a safer place than Moscow for a company that on occasion (even if unfairly) has been accused of asset stripping. No one in Jacksonville seemed particularly upset that ITERA's assets had been stripped from Gazprom nor did they seem to care that according to rumors that may have been part of a campaign of disinformation to discredit ITERA, almost all the trustees of ITERA seemed to be close relatives or mistresses of senior Gazprom executives. As of this writing, the identity of those trustees has not been published. That was nothing unusual in the Yeltsin era.

Putin's ouster of Chernomyrdin and Vyakhirev from Gazprom plus his subsequent removal in March 2002 of Viktor Gerashchenko as chairman of the Russian Central Bank were all efforts to halt such banditry and punish what Putin saw as mismanagement and the personal pillaging of state assets. (In the case of Gerashchenko, he had been accused of misusing the powers of Russia's Central Bank.) All three—Gerashchenko, Chernomyrdin, and Vyakhirev—had been long-serving state officials, and all had begun their careers in the Soviet era and had become members of the nomenclatura, the Soviet bureaucratic elite. With their removal, ITERA soon lost most of its contracts and in a short time it surrendered its position as Russia's second largest producer of natural gas to another firm, Novatek.

Putin's purge of Gusinsky and Berezovsky was of a different nature. Both were viewed as upstarts from the murkier side of the street. Unlike Chernomyrdin, Vyakhirev, and Gerashchenko, neither Gusinsky nor Berezovsky had served as a senior government official in the Soviet era. Nor were Gusinsky and Berezovsky ethnic Russians. Although Berezovsky had an advanced degree in economics, like so many upstart oligarchs he began building his wealth as a trader. Gusinsky emerged from the black market of the Soviet era—an economic criminal by Soviet standards. Berezovsky had been closely involved with criminal groups as well. Neither would ever have been allowed into the upper ranks of the Communist Party. They had not come up through the system like Chernomyrdin, Vyakhirev, and Gerashchenko. Gusinsky and Berezovsky were both at least partly Jewish and were not "ole boys" or part of the Soviet nomenclatura. Both sets of men had enriched themselves at the expense of the state, but somehow the excesses of Chernomyrdin and Vyakhirev at Gazprom and Gerashchenko at Gosbank were not regarded as venal and therefore not punishable with imprisonment or exile (apparatchiks will be apparatchiks, and besides, they are ours). Conversely, Gusinsky and Berezovsky—most definitely not "ours"—were both either imprisoned or threatened with imprisonment.

THE ATTACK ON YUKOS

This distinction about status in the Soviet era, even if subtle, also helps explain the arrest of Mikhail Khodorkovsky. The attacks on Yukos and Mikhail Khodorkovsky highlight Putin's determined effort to reign in these upstart oligarchs and at the same time renationalize and refashion their property into state companies and his vaunted national champions. By 2003, with the earlier arrests and firings, it should have been clear just what Putin was attempting to do. Yet Khodorkovsky, by his actions and his hubris, acted as if he were invulnerable. At the time, *Forbes Magazine* estimated that his net worth amounted to $15 billion, which made him the richest man in Russia. This may have warped his judgment and made him think he was indeed invulnerable.

Khodorkovksy's rise to fortune and infamy began when he was a student at the Medeleev Chemical Technical Institute. Taking advantage of the new 1987 Gorbachev-era reform authorizing the creation of private businesses, Khodorokovsky, along with twelve classmates, opened a cooperative coffee house and discotheque, which they called Menatep (this stood for Intersectoral Center of Scientific Technical Progress). They soon expanded their activities to sell consumer goods such as computers and other products that were in short supply. Trading proved to be very profitable, and all the cash they had accumulated allowed them to open their own bank the following year. This was made possible in 1987 by another Gorbachev reform that authorized the formation of private banks, the first time since the Revolution. Eventually, they named the bank Menatep, the same as their cooperative.

As one of the first private commercial banks in Russia, Menatep was in a key position in 1992 to buy up the vouchers that President Boris Yeltsin decided to issue to every Russian. The vouchers, in turn, could be exchanged for shares of stock in the thousands of heretofore state-owned enterprises that were then being privatized. The intent was to make every Russian a stockholder, a true case of people's capitalism. The hope was to involve each Russian in the privatization process and so give each one a stake in the new market system.

But as we saw, few Russians had any appreciation for the value of their vouchers. The voucher system, however, was made to order for economic sophisticates like Khodorkovsky and his banking associates who understood the potential of the vouchers. They quickly bought up as many as they could. In a few months, their vouchers enabled them to accumulate a large corporate empire.

Khodorkovsky's biggest acquisition, however, came when he managed to gain control of the oil company Yukos. This was a by-product of the Loans for Shares scheme described in Chapter 3. To help the government pay its bills, Khodorkovsky's Menatep, along with several other banks, offered to lend the government the money it needed. As collateral for its loan, Menatep agreed to take the government's stock in Yukos, a petroleum company that had been spun out of Rosneft in November 1992. Incidentally, the name "Yukos" reflected the merger of the Production Association Yuganskneftegaz (Yu) with the refinery KuybyshevnefteOrgSintez (Kos).[13] When the government could not pay back its loan, Menatep proceeded to auction off its collateral, as it was allowed to do. The assumption was that a fair auction would fetch a price high enough to provide the state with funds not only to repay the loan to Menatep but also to generate additional income for the government. What happened, of course, was that the December 8, 1995, auction was rigged. As with the other auctions, viable competitors were prevented from bidding so that the winner was, in fact, a "straw" put up by Menatep, which conducted the auction. This way, despite higher bids from Alfabank, Inkombank, and Russian Credit Bank, all of which Menatep had ruled out on a technicality, Khodorkovsky was able to pay only a bit more than $350 million, the required minimum price for control of 88 percent of Yukos stock. A few months later Yukos would have a market value of $3–5 billion.[14]

It was disturbing enough for the public to learn how these one-time state properties had been acquired by Khodorkovsky and a dozen or so other oligarchs at a fraction of their value, but to make matters worse, it happened at the same time the Russian economy was all but disintegrating. Between 1990 and 1998, as the country moved into shock therapy and simultaneously closed down much of the military-industrial complex, the GDP shrank by 40–50 percent. Not only were a few oligarchs taking out a gigantic piece of the economic pie for themselves but the pie itself had shrunk to barely half of what it had been before Yeltsin rose to power. That is why by 1998, more than one-third of the population found itself below the poverty line.

This was not the only controversial action taken by Khodorkovsky and his associates on their way to control of Yukos. Khodorkovsky and Menatep did not control all the stock in Yukos nor in the Yukos subsidiaries that produced the oil. Other investors had also purchased vouchers and exchanged them for shares, which they now owned. One of the most adventurous investors was a foreigner, Kenneth Dart, an American who was an heir to the Dart Cup business. He put in approximately $2 billion to purchase shares in Yugauskneftegaz and other subsidiaries of Yukos. Khodorkovsky wanted Dart out. Consequently, Khodorkovsky stripped Yugauskneftegaz of its value in the expectation that Dart would conclude he should sell out while there was still some value to his investment. To nudge him along, Khodorkovsky ordered that the petroleum produced by Yugauskneftegaz should be sold at below or close to cost to another subsidiary more closely controlled by Khodorkovsky. This second subsidiary would then sell the petroleum at the higher market price and thereby capture all the profit. This so-called transfer pricing is a common way to squeeze out minority shareholders (it is also a way of trying to reduce taxes for the first subsidiary) such as Dart, even if it means bringing a company like Yuganskneftegaz to the brink of bankruptcy. Subsequently, Dart initiated a series of lawsuits in an effort to recoup his investments. Despite some success, Dart estimates he lost over $1 billion.[15]

Something similar happened after Menatep closed its doors in the wake of the financial crisis of August 17, 1998. Those with deposits in Menatep as well as the foreign companies and banks that had provided loans to Menatep lost almost all their money. But while outsiders were left with little or nothing, Khodorkovsky, as we saw, transferred the bank's assets that still had value to another, but independent, subsidy in St. Petersburg called Menatep St. Petersburg.

As troubling as such behavior is, it was the physical violence, including murder, allegedly carried out by several Yukos officials that was the most disturbing. That at least was what the judges decided in March 2005 when they found Alexei Pichugin, head of Yukos security, guilty of murder and sentenced him to twenty years in prison. (In August 2007 the sentence was extended to life in prison.) In addition, Leonid Nevzlin, one of Khodorkovsky's closest associates who is now exiled in Israel, was also charged with being Pichugin's accomplice in similar crimes.[16] Given how dependent the judges in the Yukos case were on Putin, there is reason to question just how impartial the judges could be. Khodorkovsky's lawyers, in fact, claim that one of the judges would excuse herself periodically to seek advice from the Kremlin about how to rule.

While both men deny any guilt, the circumstantial evidence is hard to disregard. The Russian procurator general claims that Pichugin and Nevzlin organized an assassination attempt in 2003 on Yevgeny Rybin, then the managing director of the E Petroleum Handesges oil company. Nevzlin was also accused of ordering Pichugin to murder Sergei Kolesov and Olga Kostina; the latter had been director of the public relations department of the Moscow Mayor's office and for a time was an adviser to Khodorkovsky. Pichugin was also charged with the 2002 murder of Sergei and Olga Gorin, a businessman from Tambov and his wife, both of whom were rumored to be blackmailing Nevzlin and Valentina Korneyeva. Until her death, Korneyeva was the director of Feniks, a Russian commercial trading company.[17]

Those accusations involved not only Pichugin and Nevzlin but also Khodorkovsky. After Khodorkovsky's Menatep and his investment group won control of Yukos in 1995–1996, Vladimir Petukhov, mayor of Nefteyugansk—the city where Yuganskneftegaz, Yukos's chief producing unit, is headquartered—began to complain about the failure of the new management to pay its taxes.[18] With weak oil prices, all the petroleum companies were under enormous financial pressure. For that matter, few companies were paying their workers on time. When they were paid it was often with goods in kind, not rubles. But as we saw in Chapter 3, when we noted Mayor Petukhov's surprise that Khodorkovsky had never visited an oil field until he gained control of Yukos, the mayor was very outspoken. An oil man himself, Mayor Petukhov fought bitterly to pressure Yukos, by far the region's chief taxpayer, to pay its taxes and other bills and to refrain from massive worker layoffs. In addition, he launched a campaign to embarrass Yukos over its effort to write off a 450 billion ruble debt to the city that had accumulated before Khodorkovsky took control. On June 16, 1998, Petukhov had the audacity, not to mention poor judgment, to write a public letter to Yeltsin as well as Prime Minister Kiriyenko that criticized Yukos for its failure to pay its share of taxes to the city. He also wrote to Duma leaders charging that Yukos was guilty of criminal acts for "concealing taxes in large quantities from 1996 to 1998."[19] To dramatize the city's case, he organized a public protest outside Yuganskneftegaz's headquarters during its annual stockholders' meeting on May 27, 1998. A month later, he was shot. The prosecutors charged it was more than coincidence, and so the guilty sentence for Pichugin.

As the mayor's experience suggests, Yukos officials did not take criticism lightly. A reporter for the *Wall Street Journal* in the Moscow bureau told me he had been warned that if he knew what was good for his health, he would stop writing negative articles about Yukos. I take such reports seriously; after having read a draft of what I planned to write about Yukos in my previous book, a senior official of Yukos agreed that I had every right to publish it, but if I did, I should expect to be sued for libel. Nonetheless, I went ahead and published it. So far there has been no such suit, probably because shortly after my book was published, Khodorkovsky was arrested and jailed. Understandably, since then, he probably has been distracted from such a frivolous pursuit as a libel suit against some hapless American professor (at least my wife hopes so).

Even though it may seem like piling on a man who is serving a long sentence in jail, there probably is a case for suggesting that Menatep and those associated with it had shady reputations. As early as 1994, the CIA issued a classified report warning that "the majority of Russian banks are controlled by the dreaded Mafia." According to those who have seen the report, the only bank mentioned by name was Menatep.[20]

As sordid as all this was, Khodorkovsky and Yukos were not the only ones to have pushed the law to its limits and on occasion beyond. The struggle for control of the aluminum industry, for example, was even more violent. Nearly a dozen executives involved with aluminum finance and production encountered various forms of bodily harm. And as with Yukos, I was personally warned to avoid criticism of some of those more colorful executives in the aluminum industry. In another instance, one of the leading personalities in the aluminum business offered me a bribe. Of course, not all the oligarchs engaged in such tactics, but it was a tough time and those who were not prepared to cut a corner now and then quickly fell from power.

What makes Khodorkovsky unusual among the oligarchs is that having survived as the fittest, he suddenly decided in 1999 to embrace reform and transparency. This often happens once someone attains a hard-fought goal. "I have made it into the subdivision; now let's raise the zoning requirements." In other words, even if I acquired my property by questionable tactics, I have something of value (Yukos), and we need proper rules and regulations so no one can steal it from me.

By the time of his arrest, Khodorkovsky had become one of Russia's most outspoken supporters of good corporate governance. Undoubtedly, the fact that by 1999 oil prices had risen from $10 and were on their way to $30 a barrel and beyond was a factor in his conversion. Khodorkovsky quickly realized that despite its shady past Yukos could become a much more valuable property if oil prices kept increasing. If Yukos looked as though it had become more transparent, it might become attractive to foreign investors.

Transparency, however, did not come easily. To begin with, it required a change of cultures, something the existing Russian management would probably have been incapable of doing by itself. Therefore, in a bold move for the time and culture, Khodorkovsky decided to bring in experienced Western managers. He appointed Bruce Misamore, formerly an executive for twenty-three years at Marathon Oil and PennzEnergy, as Yukos's chief financial officer. Misamore's task was to introduce international accounting standards and bring in Western accountants. This was not easy, but when I interviewed him, Misamore insisted that whenever he met resistance from others in Yukos, Khodorkovsky provided the necessary support. Similarly, Khodorkovsky hired Steven Theede, formerly an executive at ConocoPhillips, as his chief operating officer. Going even further, he decided to staff the board of directors with foreigners and appointed Sarah Carey, a Washington lawyer; Raj Kumar Gupta, a former vice president of Phillips; Bernard

Loze of France; Jacques Kosciusko-Morizet, a former vice president of Credit Lyonnais; and Michel Soublin, the treasurer of the oil service company Schlumberger to his board. He also set up a philanthropic foundation with a blue ribbon international board, which provided grants to Russian and foreign groups, including the U.S. Library of Congress.

Khodorkovsky's embrace of transparency, while far ahead of his peers' behavior, was not beyond criticism. In 1999, for example, he suddenly relocated a stockholders' meeting without bothering to notify stockholders who were not part of management. He was also accused of stripping assets from one of his subsidiaries, the Eastern Oil Company.[21] Nonetheless by the year 2000 there were fewer such accusations, and Yukos seemed to be on its way to becoming the model of good governance in Russia. In recognition of these reforms, shares of its stock were listed on the London Stock Exchange.

Yukos also seemed to be setting high production standards.[22] Output increased by as much as 12 percent a year. Some critics complained that this was a result of over-pumping, not new exploration. In any case, by 2004, Yukos was Russia's largest producer and Khodorkovsky had become a major presence at international oil conferences. Yukos even sent oil tankers to Houston as a forerunner of what Khodorkovsky said was Yukos's willingness to become a major supplier to the United States. As part of that effort, he, along with some of the other oil company oligarchs, called for the construction of an oil pipeline to Murmansk on the Barents Sea. This would provide larger deepwater tankers with easy access to Russian oil, which would make it profitable to ship petroleum to the United States. It was, however, a direct challenge to Transneft, a state-owned company that had monopoly ownership and control of all of Russia's crude oil pipelines, including those used for crude oil exports.

Not only did Khodorkovsky decide to take on Transneft by threatening to end its monopoly in the European part of Russia but he also began a campaign to build a pipeline through Siberia to China. Yukos, on May 28, 2003, even signed a twenty-year oil-delivery contract with China. This committed Yukos to deliver 20 million tons of oil annually by 2005 and 30 million tons a year by 2010.

What arrogance. Khodorkovsky and Yukos were acting as if they were sovereign powers. Here they were, making foreign policy with China, something Putin regarded as the state's and his, not an oligarch's, prerogative. Khodorkovsky also let it be known that he was on the verge of selling off a substantial portion of Yukos to either or both Chevron and Exxon-Mobil.[23] In fact, a protocol of understanding agreeing to the sale was signed between Yukos and Exxon three weeks before Putin had Khodorkovsky arrested in October 2003.[24] An employee of Exxon has acknowledged to me that the American company had completed its due diligence study and was prepared to become a Yukos partner, much as BP had just arranged that September with TNK.

Except for a detail or two, Exxon's purchase of Yukos stock, possibly more than 50 percent of it, awaited only the signing of the contract. The CEO of Exxon, Lee Raymond, arranged to meet twice with Putin in September 2003 (once in New York and once in Moscow) to discuss the purchase. He left Putin's office under the impression that the government would not object. Early in October 2003, Raymond spoke at a business conference in Moscow on a panel with Khodorkovsky the same morning that Yukos announced its planned merger with Sibneft, the oil company Berezovsky had turned over to Roman Abramovich. Raymond and Khodorkovsky both refused to confirm news reports that Exxon was about to buy at least 40 percent of Yukos, but nonetheless officers of Exxon reported they were close to a deal.[25]

That Khodorkovsky was also close to an agreement with Chevron is confirmed indirectly by Khodorkovsky's lawyers who have filed a suit to subpoena Chevron to release its due diligence materials prepared before its intended purchase. These materials, the lawyers say, will show that Chevron had determined that Khodorkovsky had not stripped the company of its assets nor laundered money as the state contends. But this is also a fairly good indication that from Chevron's point of view, such a sale was all but ready to be made.

But while all these negotiations were under way, there were also ominous signs. Shortly after his and Lee Raymond's presentation at the Moscow conference, Khodorkovsky's wife called him in panic to report that the police had just surrounded their house and were searching documents and computers in a nearby home and a boarding school funded by Yukos. The police claimed that Yukos had donated some of its old office computers to the school and that incriminating records could be found on the hard drives. Furious that his home and family had been subject to such intimidation, Khodorkovsky called a press conference. Apparently feeling invulnerable, he dared the police to go after him. "If the goal is to drive me from the country or put me in jail, they had better put me in jail."[26] A few days later, on October 25, 2003, they took him at his word.

From Putin's point of view, Khodorkovsky was acting like a king, not a subject. In a subsequent interview reported by the *Wall Street*

Journal, Putin expressed his pique that neither Exxon nor Yukos had consulted with him in advance about such a large transaction.[27] What right, Putin implied, did Khodorkovsky have to turn over ownership and control of Russia's most valuable resource to a foreign company, and an American one at that? Evidently, Putin did not consider his September 2003 meeting with Lee Raymond to be enough advance notice. Even without such an investment by Exxon-Mobil, companies partially or substantially owned by foreign companies or investors were already producing 26 percent of Russia's oil.[28] Given historic Russian xenophobia, that was too much. The failure to consult with Putin about such a matter was not the type of respect that Putin expected from his subjects.

Having become Russia's richest man, Khodorkovsky apparently believed that he no longer needed to kowtow to political godfathers—that is, pay for a *krisha* or "roof," as Russian businessmen had since the days of the czar. Nor did he take seriously Putin's July 28, 2000, warning that if they knew what was good for them, the oligarchs would stay out of politics. In fact, Khodorkovsky seemed to think that he could create his own rival political power base. He and some of his Yukos executives became major financial supporters of several of the country's opposition parties, including the pro-Western Yabloko Party. With the help of some financial inducements, he had lined up as many as 100 members of the Duma who would support whatever he wanted. That was one of the main reasons for the defeat of two government efforts to increase taxes and environmental restraints on the oil companies. After some financial contributions, even members of the Communist Party somehow agreed to set aside their ideology and rally to Yukos's causes—causes somehow overlooked in the Communist Manifesto. There was even talk that Khodorkovsky had decided he would run for president in 2008 after Putin's term came to an end.

As if he feared no one, Khodorkovsky began to challenge not only what the Russians call the "siloviki"—the law and order types in the government who had previously served in the KGB and in other higher security posts—but Putin himself. Khodorkovsky seemed oblivious to the fact that it was particularly hard for these loyal officials to accept such a reversal of roles. In the Soviet era, they ran the country and no one dared to challenge them. Now these KGB alumni and their siloviki counterparts from similar agencies found themselves having to stand by and suffer the antics of the likes of a Khodorkovsky and his high-waymen. Who did they think they were? As one insider told me, a Yukos official close to Khodorkovksy even warned a Kremlin staff member that he would be crushed if he were to dare move ahead with an effort to increase taxes.

The conflict between Khodorkovsky and the Putin government came to a head, however, when Khodorkovsky decided to criticize Sergei Bogdanchikov, the CEO of the state-owned Rosneft. On live TV in February 2003 Khodorkovsky had the effrontery to complain to Putin that Putin's close friend Bogdanchikov had worked out a sweetheart deal at the country's expense. According to Khodorkovsky, his rival Bogdanchikov overpaid $622.6 million for Northern Oil, a company controlled by Andrei Vavilov, an insider who was a senator in the Council of Federation and a former deputy finance minister. Khodorkovsky in effect implied that Bogdanchikov and Vavilov were in cahoots with each other and had used state funds to enrich themselves. Khodorkovsky charged that Bogdanchikov paid Vavilov double what the property was worth.[29] According to Khodorkovsky, this was a corrupt kickback scheme. (Since his own background had not been so stellar, such an accusation had to be a little presumptuous coming from someone like Khodorkovsky.)

Pushing his luck even further, Khodorkovsky then told President Putin, "Your bureaucracy is made up of bribe-takers and thieves."[30] (Khodorkovsky had a point. Subsequently, Vavilov was charged in a Russian court of having committed fraud by selling shares in Northern Oil that he did not own.)[31] Defending his buddies and Rosneft, Putin insisted to the contrary. Rosneft "is a state company and needs to increase its insufficient reserves," and if anything, it is the nonstate private oil companies (Yukos) that have excessive reserves. "We still have to investigate" how they obtained them.[32]

There is good reason to believe the decision to destroy Yukos as a viable company may have been triggered by that incident. During the same month that Khodorkovsky attacked Bogdanchikov, an American working for an American investment bank in Moscow had an interview with Yury Shafranik, an insider who had been the minister of Industry and Energy from 1993 to 1996. At one point in the conversation, Shafranik became very angry because the American's firm had been recommending Yukos as an investment to its clients. "It was a mistake to promote Yukos," Shafranik warned, "because in a year's time, Yukos would no longer be in existence." Then revealing the anger that such former apparatchiks of the Soviet era have for these arriviste new owners of Russia's oil and gas wealth, he added, "Before long, there will be some real oil specialists put in charge who will know what they are doing."[33]

As Khodorkovsky became an ever more dominant and annoying presence, those around Putin came to regard Khodorkovsky as a major threat to their authority. In what is alleged to be a wiretapped conversation, the same Sergei Bogdanchikov of state-owned Rosneft (and the one Khodorkovsky accused of taking kickbacks) is heard complaining that the Yukos leader had become too uppity. "Three days in Butyrke Prison and (the Yukos leaders) will understand who is the king of the forest."[34] In Bogdanchikov's eyes, Khodorkovsky and his partners seemed to be acting as if they, not government officials, were running the country. Not only were they accusing Kremlin insiders of corruption and signing major petroleum delivery contracts with the Chinese government that had not been approved by the state, but they were also challenging the state's monopoly control of the country's petroleum pipeline. As if that were not enough, they were on the verge of selling some of the country's most valuable assets, its oil fields, to an American company. Virtually all those around Putin regarded Khodorkovsky's actions as an affront to the state's authority and agreed that it was necessary to crush Khodorkovsky as soon as possible to abort such a sale.

The counteroffensive against Yukos began with the June 2003 arrest of its security chief, Boris Pichugin, on murder and attempted murder charges and continued with the arrest of Platon Lebedev, one of the top Yukos officials, the following month. Lebedev was accused of failing to invest as much as he promised in a fertilizer company Menatep took over from the government. That may have been a violation of a contractual agreement but it hardly warranted an eight-year jail sentence.[35] Outwardly, at least, these arrests did not seem to bother Khodorkovsky. He continued to travel abroad, including a subsequent July 2003 visit to Sun Valley, Idaho, where he mixed with senior American government and business leaders, including Bill Gates, Warren Buffet, and New York Mayor Michael Bloomberg.[36] The week earlier he met with Vice President Richard Cheney to discuss Exxon-Mobil's pending offer to purchase major portions of Yukos stock. Yet despite the arrest of some of his close associates and increasing indications that he, too, might be arrested, Khodorkovsky invariably always returned to Russia. In fact, he did expect to be arrested, but he evidently assumed that he was powerful enough and his friends and lawyers influential enough to win his freedom. Khodorkovsky's lawyers have reported that he was much more alert to what was happening than he let on. At an October 11, 2003, meeting with his lawyers, just two weeks before his arrest, Khodorkovsky discussed with them the steps to be taken should he be arrested.[37]

Arrested he was on October 25, 2003. Masked police with submachine guns raided his private jet as it was refueling in Novosibirsk. He and Yukos were charged with tax evasion, grand theft, fraud, forgery, embezzlement, and extortion. He was found guilty and sent to prison in Krasnokamensk in Siberia. The initial sentence was for eight years, but additional accusations made in 2007 could result in another fifteen years.[38]

In his frenzy to punish Khodorkovsky and destroy Yukos, the prosecutor general not only arrested Yukos's senior executives but he also went after anyone remotely associated with Yukos, including its lawyers and junior staff. While the list is incomplete, Table 5.3 lists over two dozen people associated with Yukos who have either been jailed or have fled into exile. The case of Svetlana Bakhmina stands out. The mother of two little boys aged two and six at the time of her arrest, she was only a middle-ranking member of Yukos's legal department. She was taken from her home at 5:00 in the morning, refused bail, and prevented from seeing her children for more than sixteen months. She was found guilty of embezzling $290 million from a Yukos subsidiary and sentenced to seven years in a maximum security prison, a sentence almost as long as Khodorkovsky's. If her sentence had been one year less, she would have been freed from prison under an amnesty issued for mothers sentenced to six years or less. In her defense, her lawyers argued that as she lived in a two-room apartment, the charges that she had stolen so much money were absurd.[39] The real reason for her arrest, claim her lawyers, was that the prosecutor was holding her hostage to force her former boss, the general counsel of Yukos, Dmitry Gololobov, to return to Moscow from his London exile.

This was typical of the way the courts treated the Yukos defendants. Undoubtedly many of the charges, at least against the company's senior executives, had some merit, but the courts' assault on mid-level Yukos employees was needlessly harsh. Like Bakhmina, most of those charged were denied bail. By comparison, those charged in the Enron case in Texas were allowed out on bail until they were found guilty. For that matter, Khodorkovsky's lawyers argued that under the Russian Constitution, "pretrial detention is basically not permitted in a white-collar trial."[40] Moreover, Yukos lawyers were also harassed, their homes and offices raided, and evidence withheld.

Khodorkovsky's lawyers have complained that there was remarkably little effort by either the judges or the prosecutor general to adhere to legal precedents or procedures.[41] Khodorkovsky's lawyers also pointed out that some of the charges against Lebedev and Khodorkovsky in the

TABLE 5.3 Yukos Senior Executives, in-house or outside counsel and accountants; left the country or arrested, in prison or under house arrest or wanted for questioning

Name	Position	Status
Alexei Pichugin	Yukos head of security	Arrested June 19, 2003. Convicted March 30, 2005 of murder and attempted murder and sentenced to 20 years in jail.
Platon Lebedev	Menatep chairman	Arrested July 2, 2003.
Mikhail Khodorkovsky	Yukos CEO	Arrested Oct. 25, 2003.
Alexei Kurtsin	Yukos executive	Arrested Nov. 18, 2004.
Svetlana Bakhmina	Yukos deputy general counsel	Arrested Dec. 7, 2004.
Vladimir Malakhovsky	Yukos trading company executive	Arrested Dec. 10, 2004.
Vladimir Pereverzin	Former deputy director of Yukos external debt department	Arrested Dec. 18, 2004.
Dmitry Velichko	President of ZAO Rosinkor	Arrested January 2005.
Oleg Vitka	Head of Yukos joint venture with Hungary's MOL.	Arrested March 10, 2005.
Yelena Guryanova	Wife of Pavel Ivlev, deputy managing partner of Yukos' law firm. Living outside Russia.	
Andrey Krainov	Director, AOZT Volna	
Steven Theede	CEO of Yukos. Wanted for questioning. In London.	

(continued)

TABLE 5·3   (continued)

Name	Position	Status
Bruce Misamore	CFO of Yukos. Wanted for questioning. In Houston.	
Mikhail Brudno	Menatep. Wanted by Interpol. In Israel.	
Vladimir Dubov	Menatep. Wanted by Interpol. In Israel.	
Leonid Nevzlin	Menatep. Wanted by Interpol. In Israel.	
Dmitri Gololobov	Former head of Yukos legal affairs in London. Svetlana Bakhmina's boss.	
Vasily Shaknovsky		Paid back taxes. Received one-year sentence. Sentence suspended.
Elena Agranovskaia	Outside counsel	Arrested on Dec. 8, 2004, on charges of money laundering and evasion.
Igor Malakhovsky	Director of Ratibor	Arrested June 10, 2005, on charges of money laundering.
Antonio Valdes Garcia	Director, Fargoil, Yukos Trading Facility	Arrested July 19, 2005. Pumped more oil than authorized.
Yury Beilin	Deputy CEO of Yukos	Ordered to be arrested, Feb. 28, 2003—exiled in London.
Mikhail Yelfimov	Acting president for refining and marketing, Yukos	
Elena Marochkina	Chief accountant Yukos subsidiary	Accused of tax evasion.
Sergei Shimkevich	Director General Tomskneft	Accused of embezzlement.
Pavel Anisimov	Executive manager Samaraneftegaz	Accused of tax evasion.

Ramil Burganov	Eastern Oil Company	Accused of theft of more than 1 billion rubles; granted political asylum in Britain.
Dmitri Maruev	Yukos chief accountant	Accused of $2.6 bill. fraud. In exile in London.
Natalya Chernisheva	Rosprom	Accused of fraud. In exile in London.
Tagirzian Gilmanov	Duma deputy. Former managing director Yuganskneftegaz	Tax evasion. Sentenced to three years in jail.
Vasily Alexsanyan	Chief of Yukos legal department	In jail, accused of embezzlement, tax evasion, and money laundering. Suffering from lymphoma.
Aleksandr Temerko	Board member, Yukos	Accused of fraud. International arrest warrant issued for his arrest.
Mikhail Trushin	Yukos 1st VP	Accused of theft.
Ludmila Siyusareva	General accountant Yuganskneftegaz	Tax evasion. Arrest warrant issued.
Igor Babenko	Manager Menatep	Charged with theft of $12 million. Asylum in Lithuania.
Oleg Grudin	Former general director Samaranefteproduct (1999–2006)	Suspected by prosecutors of failing to pay 1.641 million rubles in property tax (2004).

Source: www.khodorkovsky.info/timeline/13425o.html

fertilizer company case had been heard eighteen months earlier at a civil hearing. After negotiations the dispute was resolved to the satisfaction of the prosecutor general, who subsequently reversed his decision. To bring up the issue again was a form of double jeopardy.

THE PLUNDERING OF YUKOS

As for Yukos itself, it was treated just as harshly. By the end of the trial it was accused of failing to pay $33 billion in back taxes.[42] Undoubtedly Yukos owed money for back taxes unpaid in its earliest years, but the government seemed unbothered by the fact that in some years the alleged tax bill exceeded total revenue earned.[43] As punishment, in December 2004 Yukos's most valuable asset, Yuganskneftegaz, was sold at yet another rigged auction for $9.35 billion to an obscure and heretofore unknown entity called the Baikal Finance Group, which turned out to be a front for Rosneft. Several outside appraisers insisted that the price was less than half of what such an auction should have yielded.[44]

To protect themselves and prevent further plundering of Yukos, in December 2004, some of the American executives working for Yukos sought Chapter 11 bankruptcy protection in a Texas federal bankruptcy court. This action frightened off some prospective purchasers, including Gazprom, which feared that if Gazprom seized Yukos property, their assets might in turn be seized by European or American courts. Hence the sudden appearance of the Baikal Finance Group, which was used to launder the assets so that they could then be sold to Rosneft. If Rosneft itself had entered such a bid at the auction, it might also have been threatened with a lawsuit in a European court. Ever since, both Gazprom and Rosneft have used straws to do the initial bidding on properties being stripped off of Yukos: Neft Aktiv for Rosneft and Unitex for Gazprom.[45]

There was reason for such fears. In 1993, Noga, a Swiss trading company, had used a Western court judgment against Russian authorities to seize a Russian bank operating in Luxembourg. It did this to force payment of the overdue bills claimed by the Swiss firm. (I had served as an adviser to Noga in its effort to collect what it claimed it was owed. Noga was unaware that Russia had such assets in the West. Once alerted, they sought support from a European Court and the East-West United Bank in Luxembourg was seized as a form of collateral to force repayment.) There was every reason to believe if Gazprom or Rosneft themselves tried to buy up Yukos assets, they would suffer a similar fate. That was enough for Gazprom and Rosneft to refrain from making a direct bid. But after due consideration as to whether a U.S. court had jurisdiction in such a case, the Texas judge decided it did not.

The company's fate therefore was left up to Russian courts. After seeing that Yukos could not pay its debts, the Moscow Arbitration Court on August 2, 2006, declared Yukos bankrupt.[46] This was despite Putin's earlier insistence at a RIA Novosti press agency discussion that Russia would not force Yukos into bankruptcy.

In 2007, a second batch of nearly 200 properties, including the 20 percent stake in Gazpromneft (formerly Sibneft) that Yukos still owned as well as its Tomskneft and Samaraneftegaz subsidiaries, were put up for sale to cover the rest of the back tax bill.[47] Because of the increase in petroleum prices after 2004, the remaining Yukos properties were appraised at a higher level and valued at a minimum of $22 billion, and then again increased to $26 billion.[48] At that appraisal, what remained of Yukos would have allowed it to sell off some subsidiaries, use the funds to pay back the taxes, and still retain at least some assets for itself. That would mean that it really was not bankrupt. To prevent that from happening, the state authorities recalculated the taxes owed by Yukos and found that they had underestimated what Yukos owed. According to their new calculation, Yukos's total debt was in fact $26.6 billion, an amount that again conveniently exceeded the company's newly estimated net worth.[49] Former Yukos executives argued, however, that the real value of Yukos's remaining assets was even higher at $37 billion. Whatever the fair value of the property, there was no way the former owners of Yukos would have been able to keep any of the assets. Whenever someone came up with a calculation that showed that the company's assets were actually more valuable than indicated in earlier estimates, the tax authorities would recalculate the overdue tax debt and somehow come out with an even larger estimate.

The charade continued. In late March 2007, Rosneft, which along with the state was the chief creditor for Yukos properties, organized an auction to sell off the assets it had seized from Yukos. The first auction involved the sale of 9.44 percent of Rosneft's own stock that Yukos had acquired earlier. Reminiscent of the Loan for Shares auctions of the mid-1990s, Rosneft was not only the auctioneer but—to no one's surprise—also the winner.

To avoid a repetition of such "done deals," after the Loans for Shares fiasco, a new requirement was put in place. All future auctions would be valid only if there were at least two bidders. Therefore, in order to lend credence to the March 2007 auction, TNK-BP agreed to enter a bid. However, TNK-BP dropped out after the opening bid raised the $7.5

billion asking price by a mere $97 million.[50] The $7.5 billion was $900 million less than it would have cost to buy an equivalent amount of Rosneft stock on the London Stock Exchange. If TNK-BP had really been a serious bidder, they would not have let such a bargain pass. It was generally agreed that TNK-BP had entered a bid only to make the exchange look legitimate and so curry favor with Putin and the Kremlin in hope that neither the state nor Gazprom would seize BP's holdings in Kovykta. There was further evidence that the auction was a charade: the acting head of Yukos in Russia at the time, Eduard K. Rebgun, who earlier had agreed to discount the price of the Yukos assets he was supposed to be protecting, applied to join Rosneft's board of directors, a blatant conflict of interest.[51] Without apparently being aware of the irony in what he was saying, after the auction Rebgun boasted that the auction "went, thank God, normally."[52] For those with memories of Loans for Shares, it certainly did. Once again, however, the sale failed to generate enough to pay off what remained of the original $25 billion tax bill.

But this was not the end of the struggle. In May 2007, the state put the Yukos headquarters up for sale. An obscure company called Prana bid an amazing $4 billion. By doing so, the total collected by the state exceeded the amount the state was owed. No matter, Yukos was still treated as a bankrupt company.

In an effort to salvage something more from the bankruptcy, some former Yukos stockholders have threatened lawsuits against what they claim has been the illegal expropriation of Yukos. The lawsuit was intended to deter Western companies from bidding and Western banks from lending money to Rosneft so they would not have the funds they needed to bid for Yukos properties. It is also why Rosneft itself did not bid in the auction. Instead, yet another straw called RN-Razvitiye, patterned on the Baikal Finance Group model, did the bidding and won. This was to protect Rosneft from a Western lawsuit. Like Baikal Finance Group, RN-Razvitiye came out of nowhere in January 2007, three months before the auction.[53] It clearly was created just for the purpose of bidding on Yukos. It turns out that at the time that RN-Razvitiye was capitalized at only $385. Nonetheless, it easily and quickly managed to borrow the $9 billion it needed to bid in the auction, something I wish my banker would let me do.

Taking no chances, Russian authorities also sought to discredit Yukos's financial reports. It sued PricewaterhouseCoopers (PWC), Yukos's accounting firm, claiming that PWC misrepresented Yukos's earnings. Initially a lower Russian court found PWC guilty and ordered that it pay $15 million in back taxes. Under pressure PWC did the state's bidding and disassociated itself from ten years of its own audits of Yukos (not something to be proud of). This move undermined Yukos's argument that it had fully paid its taxes.[54]

While the Yukos case has to be an embarrassment for those who insist that Russia is "a normal country" that is coming to embrace the concept of "rule of law," it did accomplish what Putin wanted—the removal of Khodorkovsky from the economic and political stage, the dismemberment of Yukos, and its effective renationalization and redistribution to state-controlled entities like Rosneft. More than that, by transferring ownership of most of Yukos to Rosneft, Putin prevented Yukos and its petroleum reserves from falling in the hands of Exxon-Mobil or any other foreign entity. At the same time, Putin enhanced the role and capabilities of another one of his designated national champions by putting another 11 percent of the country's petroleum output back in the state's hands.[55]

SIBNEFT'S TURN

Having reined in Yukos, Putin's next target was Sibneft. Its takeover was done so quietly hardly anyone noticed. As we saw, when threatened with jail, then chief owner Boris Berezovsky hastily fled to London. On his way out, he sold off his share of various properties to his one-time protégé, junior partner, and erstwhile sportsman Roman Abramovich, who in turn was happy to relay most of his new possessions on to the state. Among other properties, Berezovsky transferred control to Abramovich of the aluminum company Rusal, the TV network ORT, and, most important, the petroleum company Sibneft. This undid an earlier merger between Yukos and Sibneft agreed to in April 2003. At that time, Khodorkovsky had paid $3 billion for a 26 percent share of Sibneft stock. After Khodorkovsky's arrest, Abramovich put the merger on hold but neglected to refund the $3 billion.

For a time, Abramovich also flirted with the subversive idea of selling half of his share of Sibneft to a foreign company. He considered offers from Chevron-Texaco, Shell, and Total.[56] But after the inevitable visits from the Russian tax authorities and significant claims of some $1.4 billion in tax arrears, in September 2005 Abramovich agreed instead to sell his 72 percent stake in Sibneft to Gazprom for $13 billion.[57] Renamed Gazpromneft (Gas Industry Petroleum), this gave state-dominated Gazprom a major stake in the petroleum sector for the first time. With the transfer of Sibneft to state ownership, the state once again gained control of 30 percent of Russia's total oil output.[58]

Don't shed any tears for Abramovich. By working with the state and Putin, he was able to sell Sibneft at a price that made him Russia's richest man with a net worth estimated by *Forbes Magazine* at over $18 billion. This leapfrogged him over the now jailed Khodorkovsky, whose earlier $15 billion had been mostly in Yukos stock. After the company's bankruptcy, it was worth only a fraction of what it had been.

Abramovich was not only Russia's richest man. Since he had effectively moved to London, he also became *Great Britain's* richest man. One of his new homes was a 440-acre estate in Sussex, the other a mansion in London's exclusive Belgravia district. His other purchases included two of the world's largest yachts and the Chelsea soccer team for which he paid $250 million—not bad for a poor boy with a murky background who started out in 1996 as a junior oil trader and office manager at Sibneft.[59]

FRIENDLY PERSUASION

No company can assume it is immune from harassment by the state, particularly if that company fits in nicely with Putin's notion of national champion. Mikhail Gutseriev, for example, who formed Russneft in 2002 by building on assets acquired from Slavneft, until then a state company, claimed that because of pressure from the police and federal tax authorites, he was being forced to sell his company to Oleg Deripaska. It was clear Deripaska was acting on behalf of Putin. At the time in 2007, Russneft had become Russia's seventh largest oil company, producing 16 million tons of crude oil a year. Gutseriev later denied that he had come under pressure but there were rumors that he had angered Putin when Gutseriev attempted to buy up Yukos assets, thereby making it hard for state-owned Rosneft to do so.[60] Other companies that have in effect been similarly renationalized include VSMPO-AVISMA, a titanium producer that was purchased by state-owned Rosoboronexport, a military arms exporter, and Nortgaz, all of whom have been forced to become national champions.[61]

With state ownership restored over what used to be Yukos and Sibneft, among the larger companies, the state lacked a majority share of the stock only in Surgutneftegaz, LUKoil, and TNK/BP. As for Surgutneftegaz, this was not a problem. For all intents and purposes, Surgutneftegaz behaved almost as if it were under state control, and as such, a national champion. Rarely if ever did the company deviate from state policy or face complaints or even allegations about law violations, environmental abuse, or tax delinquency. The CEO, Vladimir Bogdanov, was an "ole boy" uninterested in the high life who preferred living in Siberia rather than in that fleshpot Moscow. He also saw nothing wrong in keeping his corporate operations as secretive and opaque as possible. Bogdanov and his staff controlled more than 50 percent of the company's stock, which they held very closely.[62]

This close control does not mean that Surgutneftegaz has managed to avoid all controversy. After being praised in 1998 and 2004 for the way its workers were treated (this was in sharp contrast to those working for Yukos, who complained about being underpaid and harassed), those same workers at Surgutneftegaz nonetheless went to the streets in a May Day protest in 2006 to demand higher wages and an end of the arbitrary awarding of bonuses to management favorites.[63] Similarly, at the other end of the ownership pyramid, outside minority stockholders have complained about insider self-dealing and the lack of transparency. The Harvard Management Company, which manages Harvard University's $34.9 billion endowment (as of June 2007), filed a claim against Surgutneftegaz with the American Arbitration Association in New York. It charged that Surgutneftegaz's management withheld $400 million in dividends that were hidden by senior management but should have been paid to minority shareholders, including $3.7 million which it said was owed to Harvard.[64] That publicity and subsequent legislation by the Russian government induced Surgutneftegaz to pay out an acceptable dividend return to its stockholders, although as of early 2007 it had not agreed to make up for the missed dividends of the past.[65]

LUKoil's willingness to adapt to state policy was not quite as swift as Surgutneftegaz's, but only rarely has it or its CEO, Vagit Alekperov, overtly challenged the state or Putin's policies. How LUKoil and the state have generally worked in tandem was highlighted in September 2003 when President Putin joined with Alekperov to inaugurate a LUKoil filling station in Manhattan. Why not? Establishing Russian beachheads in the center of New York City and New Jersey is just what "national champions" are supposed to do, whether they be state or privately owned. Like any good multinational corporation, LUKoil, as we mentioned earlier, bought into the U.S. market by acquiring Getty Petroleum Marketing Limited in 2000. This brought it 2,000 stations stretching from Maine to Virginia, and it bought another 1,000 from Mobil. It expects all these stations to be converted into LUKoil outlets by 2008. This will give LUKoil 7 percent of the market in its territory but as much as 24 percent in New Jersey and Pennsylvania (where it is concentrated).[66] Initially, these stations were supplied with crude oil from the Middle East, which was then refined in the United States. LUKoil plans, however, to replace that Middle East oil with its own crude from Russia in 2008. Reducing

U.S. reliance on Middle East oil would be very much in its national interest, but given how Russia uses its petroleum exports for political purposes, there is no guarantee that future imports from Russia will not at some time be used in a similar way.

In addition to its retail presence in the United States, LUKoil, as a national champion, also purchased from ConocoPhillips 376 Jetts gas stations that were located in Belgium, the Czech Republic, Slovakia, Poland, Hungary, and Finland. This is in addition to the refineries it operates—or plans to operate—in Turkey, Kazakhstan, the Netherlands, Bulgaria, Romania, and Ukraine. Before the U.S. invasion, LUKoil also operated in Iraq.

In all these endeavors, LUKoil consulted with President Putin. It was also careful to ask for Putin's blessing when LUKoil wanted permission for ConocoPhillips to buy up a substantial percentage of LUKoil stock. Putin agreed but insisted that ConocoPhillips limit itself to no more than 20 percent ownership. He seemed comfortable with an investment of that size. Unlike the TNK-BP arrangement, holding ConocoPhillips to only a 20 percent share was more likely to ensure that LUKoil would retain operating control.

IS TNK-BP NEXT?

Putin seemed especially appreciative that LUKoil had the courtesy to ask him for his approval before negotiations were completed, not after, as was apparently the case when BP bought 50 percent of Tyumen Oil (TNK). While Putin attended the TNK-BP merger ceremonies along with British Prime Minister Tony Blair in London, it later became clear that Putin was not happy with this arrangement. TNK made the mistake of not only involving Putin after most of the terms had been agreed to but additionally of allowing BP to take over the company's management and thus gaining effective control of a valued Russian resource. Officially this was a 50/50 partnership, but under the agreement, BP personnel took over operating control of the partnership.

Nevertheless, despite some rough patches when the BP personnel assigned to the partnership found their work culture at odds with TNK's, BP was able to bring a more efficient and productive management to the operation. Using its own technology, BP has been able to tap deposits heretofore beyond TNK's ability to exploit; as a result, it discovered that TNK-BP has considerably more workable reserves than TNK realized it had.[67]

By mid-2007 there were recurring rumors that Putin and the Russian government were not happy with the extent of BP's involvement and that the government was seeking some way to ease the private Russian owners of TNK out of the partnership, put state-controlled Gazprom in their place, and then reduce BP to a minority stockholder. Putin and Co. may again be maneuvering to create another national champion and in effect, renationalize the company.[68]

In hopes of preventing pressure on BP, the CEO of BP, Lord John Browne, and his successor, Tony Hayward, sought to ingratiate themselves and BP with Putin. That is why they arranged a meeting with President Putin in March 2007 and proposed that BP bid in the auction to buy 9.44 percent of the Rosneft shares that Yukos owned before it went into bankruptcy. As we saw, to be legal, there had to be at least two bidders in the auction and at the time it looked as though there would be only one bidder, Rosneft. By entering its bid and thereby ensuring that there were two bidders, but not bidding enough to win, BP acted to help out Rosneft and Putin. For much the same reason BP bought up $1 billion of Rosneft stock in an initial public offering of its stock in London earlier in 2006.[69] By doing so, BP pushed up the price of Rosneft stock, and yet its stock purchase was not large enough to give BP any operational control. BP hoped that these two gestures combined would ward off future attempts by Gazprom or even Rosneft to muscle out either their Russian oligarchs or BP itself from their TNK-BP partnership.

All of this chess board maneuvering has led to a good deal of second guessing as to whether there would have been less harassment of TNK-BP if BP had settled in the beginning for just 49 percent ownership. This became clear after Putin declared that exploration in new Russian fields off and onshore would be limited to companies in which the Russian partner had at least 51 percent control. Given the historic reluctance in Russia to let foreigners have too much control over Russian resources (foreign money is welcome; it is the control that is not), Putin might well have set the Russian share even higher if BP had agreed to only 49 percent or even 45 percent. Compared to ConocoPhillips's 20 percent limit, anything even close to 50 percent was considered dangerous.

IT'S TIME TO CHANGE PARTNERS AGAIN

With petroleum prices hovering around $100 a barrel, all of Russia's petroleum producers have prospered. Nevertheless the companies that are not state-dominated national champions, even those firms controlled by Russian executives, are occasionally reminded that they are there to do the state's bidding. Like a well-bred and carefully trained horse that still needs the periodic sting of the whip to remind the horse that it is a horse and the man in control is the jockey, the Kremlin will almost as a matter of routine periodically send in tax collectors and inspectors, not only to collect taxes and carry out inspections but also to harass. Beginning in 2006, environment authorities joined in this minuet. Almost every private energy company operating in Russia (foreign and Russian alike) has been subjected to such visits and harassment in one form or another. LUKoil, for example, has had to deal with charges that it was behind schedule in exploration, drilling, and starting production at eleven oil fields in the Komi Republic.[70]

As often as not, given the sorry track record of Soviet and the successor Russian oil companies in polluting their oil fields, there is probably something to the pollution charges. But recently there has often been another reason for these accusations. The unpublicized agenda for such warnings is meant to mask the effort by state-dominated Rosneft or Gazprom to muscle their way to an equity position in these private ventures at a reduced price. That seemed to be the real motive in a whole series of cases: LUKoil in the Komi Republic; Royal Dutch Shell, Mitsui, and Mitsubishi on their Sakhalin II project; Exxon and its partners on the Sakhalin I project; as well as threats against the TNK-BP partnership at the Kovykta natural gas project near Irkutsk, a threat the BP March 2007 bid for Rosneft stock seemed designed to ward off.[71] Once Gazprom becomes a partner, especially if it becomes the dominant partner, the charges of pollution miraculously seem to disappear.

SAKHALIN

The Shell dispute in Sakhalin is complicated. Sakhalin is a large island off the Russian mainland in the Sea of Okhotsk north of Japan. There was evidence as early as the late nineteenth century that Sakhalin had deposits of oil. But because the environment there involves such extremes in weather and offshore working conditions, Soviet and then Russian companies were unable to work the deposits on their own. Thus in 1975, Soviet authorities agreed to allow Japanese companies in to explore the region for gas and oil. This was one of the rare instances after World War II when the Soviet Union allowed a foreign company to engage in commercial activity inside the Communist state.

Working conditions on and around Sakhalin are some of the most challenging in the world. Russian authorities understood that and knew that most of the work would have to be done offshore. It is so cold that the sea freezes over most of the winter, putting a halt to existing work; and the ice floes are a continuing risk to the drilling rigs. That was why Russian authorities concluded they could not do the work themselves and agreed to sign favorable Production Sharing Agreements (PSA) with foreign companies who have had more experience working in such a difficult environment. Exxon-Mobil, for example, solved the ice floe problem by locating its drilling rig on dry land and then after drilling vertically, it redirected its drilling efforts horizontally under the sea to the oil-bearing deposits. That is a technology that Russian companies have so far not been able to master.

Yet it is easy to understand why the Russian authorities went after Shell to revise the original Production Sharing Agreement (PSA). Shell had initially indicated that developing their project would cost $10 billion. That allowed for how risky the work would be. Under the terms of the PSA, only after Shell and its partners (none of whom were Russian companies) had recouped their costs would Russia begin to share in the profits of the operation. In July 2005, however, Shell announced that it had underestimated the costs and challenge. In fact, because of higher steel prices and more complicated work, the cost would be $20–22 billion.[72] About the same time Exxon reported that its costs would also be higher than anticipated, not the originally estimated $12.8 billion, but $17 billion. This was high but not double the original price like Shell's.[73] But if the Russian government were to wait while Shell recouped $20–22 billion, they might never see any profit. Whether the higher estimate was accurate or not, Shell must have known the higher cost would upset the Russians. The Russians could understand that there might be some increase, but not as high as twofold.

Not surprisingly, therefore, the Russians began to pressure for a rewriting of the original PSA. They would have done that as a matter of course even if Shell's costs had not risen so much. As we have seen, concessions such as PSAs offered at a time of need tend to be disavowed once Russia regains its strength and self-confidence.

As in the past, the Russian government sought to protect its interests by forcing Shell to include Gazprom in the venture as a partner. Remember that Sakhalin II was the only PSA project until that time that had no Russian partner. Recognizing its shaky status, Shell agreed to yield to Gazprom and sold it a 50 percent equity plus one share at the bargain price of $7.45 billion. Considering Shell's earlier estimate that the project would entail an expenditure of $22 billion, Gazprom evidently was able to bully its way in for $3.55 billion less than it should have paid for a half interest share. As the Godfather would have said, Gazprom made Shell an offer that it couldn't refuse. What was embarrassing to outside observers, however, was the alacrity, even enthusiasm, with which Shell, Mitsui, and Mitsubishi agreed to Gazprom's offer, as if they normally write off $3.55 billion every day. Insisting that Shell harbored no hard feelings, the CEO of Shell, Jeroen van der Veer, fairly bubbled over with gratitude to Gazprom for its willingness to step in as a partner for such a trifle while he also "enthusiastically thanked Mr. Putin for his support."[74]

Shell is not the only company that has been forced or found it necessary to kowtow to the Russians. Total, the French petroleum company, has been equally submissive and humble. Like all energy companies, it has discovered that as-yet-untapped investment opportunities are more and more difficult to find, so it must take what is offered. A good example is how Total responded after Gazprom changed its mind and decided to bring in Western companies to help it develop the vast but difficult to work Shtokman natural gas deposits. Earlier Gazprom had solicited proposals from several Western energy companies as to how they would develop the Shtokman gas fields, but in October 2006, Gazprom rejected them all and decided it would do the work itself. However, after reflecting on the location of the deposits—500 to 600 kilometers (300–360 miles) offshore in the Barents Sea with its icebergs and storms—Gazprom decided to seek Western help after all. It selected Total from a half dozen companies that offered to do the work even though Total has relatively little experience in such harsh Arctic work conditions. Total does, however, have extensive experience with liquified natural gas (LNG) operations, and much of the Shtokman gas will eventually be shipped in liquified form.

There was little doubt that Total was eager to win the contract. That explains why it agreed to participate even though it will have no equity in the project. Total insists this will not prevent it from carrying some of the Shtokman reserves on its financial statements, something all energy companies are under pressure to do because the more reserves that are listed on their books, the higher the price of the company's stock is likely to be. The reason it may not be able to include some of the Shtokman reserves on its books is that Total has agreed to operate primarily as a service company; in addition, the project involves enormous risk. But since it wanted to be involved, Total did not have much choice in the matter. This was another instance where the Russians realized that they could drive a hard bargain—and they did.

Total is not alone. Exxon and its partners in their PSA in Sakhalin I came under similar pressure. Strictly speaking, the Russian authorities did not question the earlier tax concession and cost estimates they had originally agreed to in December 1993.[75] That would have been an outright contract violation. Instead, they latched on to cases when both Shell and Exxon had violated pollution standards (in some instances, serious violations) as a way of calling for the cancellation of the original PSA.

Gazprom used the same tactics on TNK-BP in Kovytka, also in northern Siberia. For fear of being pushed out of Kovytka entirely, BP offered Gazprom a controlling share in the project, which according to some estimates is worth as much as $20 billion. For BP's 62.42 percent stake in Rusia Petroleum (worth about $12.5 billion), which holds the license to develop Kovykto, Gazprom has agreed to pay between $700 and $900 million—quite a bargain, at least from Gazprom's point of view.[76] It did not seem to matter that the reason for TNK-BP's failure to produce the 9 billion cubic meters of gas per year it had promised is that Gazprom refused TNK-BP access to its monopoly pipeline network. The only alternative was for BP to sell its gas to a nearby community. But because there was so little industrial development there, there was only need, at most, for 2.5 billion cubic meters of gas in the region.[77] If it were to produce 9 billion cubic meters of gas as it had promised, the only thing it could do with it would be to burn it (flare it). That would not only be a waste of a valuable resource but would violate a Russian law and also add to the carbon dioxide in the atmosphere, a form of pollution that would also warrant criticism. To Putin, BP's failure to act was inexcusable. In a June 1, 2007, press conference, Putin pointedly insisted that BP had been fully aware of these requirements beforehand and should never have entered an agreement if it could not meet them. The Russian owners of TNK were reported to believe that this was all a pressure tactic to force them to sell their share of the partnership at a cheaper price to Gazprom.[78]

There was also speculation reported in *Forbes Magazine* online that in an effort to hold on to its stake at Kovykta in the north, BP had offered to create a joint venture with Gazprom that would provide Gazprom with an equity interest in BP's LNG operation in Trinidad and Tobago.[79] Such an offer would require that Gazprom allow BP to stay in Kovykta at least as a partner with Gazprom. What makes this attractive to Gazprom is that the Trinidad-Tobago facility that BP operates there is the supplier of 65.5 percent (16.56 billion cubic meters) of the LNG the United States consumes. If such a joint venture is created, it will provide Gazprom with its first major entry into the U.S. natural gas market, something they can broach at this point only with LNG capability.

Gazprom's refusal to allow petroleum producers to ship their by-product gas through the Gazprom pipeline network was a legacy of the Soviet era when the Ministry of Gas was assigned a target in cubic meters of natural gas and the Ministry of Petroleum was assigned a target in tons of petroleum. Neither ministry was credited if it produced the other's product. For the Ministry of Petroleum, the easiest way for its producing units to dispose of the by-product gas released as they extracted crude oil was to flare it. When these oil wells were privatized, the private firms saw the value in the by-product gas, and it made as much sense for them to burn money as to burn gas. By contrast, Gazprom, even though it has private shareholders, is still essentially a state-dominated company that has not fully rid itself of the Soviet bureaucratic culture, and so profit is not the only or even uppermost consideration. And since these "Gazoviki," as John Grace says they are called,[80] control the major cross-country pipelines, energy producers have to play by their rules, which strictly limit the amount of natural gas produced by non-Gazprom-controlled units into the Gazprom pipeline distribution system. That is why, according to a report in the *Moscow Times* and estimates by the World Bank and the International Energy Agency, Russia accounts for almost 11 percent of the more than 110 billion cubic meters of gas flared worldwide each year into the atmosphere.[81]

As if to show they play no favorites, not only did the Russian government harass TNK-BP and Shell, which are British and Dutch, and LUKoil and its minority stockholder, ConocoPhillips, which is American; they also went after the French company Total in 2006. It, too, had been granted a Production Sharing Agreement in December 1995 to develop the challenging Khargyaga oil project in the Nenets Autonomous District in the far north. Just as with similar projects in the Russian Far North, the weather is extreme: bitter cold and dark in the winter and swampy and infested with mosquitoes in the summer. The mid-1990s was also a time when the Russian economy had serious problems and needed all the outside help it could get. Total had a 50 percent share in the project. Forty percent of the remainder was held by Norsk Hydro of Norway, and 10 percent by the Nenets Oil Company, which is owned by the Nenets Autonomous Region. Total was charged with failing to drill as many wells as it had promised. Moreover, Total also failed to pump the associated gas released with the crude oil from the well back into the well. Instead, it flared that gas. As a penalty, Total was told its license for the PSA would be withdrawn.[82]

Such threats should not have come as a complete surprise to Total. This was not the first time the rug had been pulled out from under an agreement or pending agreement originally made at a time when Russia was relatively weak. In September 2004, Total had all but concluded a deal to invest $1 billion in Novatek, a semi-private gas producer in Russia.[83] According to Total, the deal was canceled after Russian regulators imposed numerous obstacles. Total attributed the cancellation of the deal to pressure from Gazprom, which wanted to exclude foreign equity investors from the gas sector.

IT'S NOT WHAT YOU DO, BUT HOW YOU DO IT

What emerges from these cases is that once they were able to revitalize their energy sector, the Russians ceased to be a supplicant. They no longer felt the need to offer the generous terms that come with a PSA—a colonial treaty, as Putin now calls it. That change in status also led Putin and those around him to find ways to regain control over the mineral assets, energy, and metals that had slipped from state control in the Yeltsin era.

In some cases, this was done by effectively renationalizing the properties; in other cases, it was done indirectly with threats of legal action as well as not-so-friendly visits from the tax authorities. Rather than a threat, sometimes all that was needed was a friendly chat. Whichever method Putin chose to follow, by 2008 and the end of his term, President Putin had effectively reversed the process of privatization, at least among what Lenin had called the "commanding heights" of Russian industry (see Table 5.4).

Putin noted in our September 2005 Valdai Hills Discussion Group meeting (organized by the RIA Novosti press agency for foreign specialists) that while we in the West have criticized the Russian government when it sought to reassert control over its energy assets, this, after all, is the pattern of ownership in all but a few countries, such as the United States and the UK.

The Western response to Putin's effort to restore the government's control over the commanding heights of Russian industry, should not be anger that the state wants to take control but with the way the state

TABLE 5.4 Renationalization and Control by Siloviki

	Renationalized	New owner	State's share (%)
Yugansneftegaz	Dec. 2004	Rosneft	100% (will be reduced to 70% after IPO)
Sibneft Oil	Oct. 2005	Gazprom	51%
AvtoVaz Automobile	Nov. 2005	Rosoboronexport	2% shares (effective control)
Kamaz Diesel Trucks	March 2006	Rosoboronexport	Already 100%
VSMPO-Avisma Titanium		Rosoboronexport	Under pressure
Gorbunov-Kazan Aircraft	Feb. 2006	United Aircraft	75%
MIG Aircraft	Feb. 2006	United Aircraft	
Sukhoi Aviation	Feb. 2006	United Aircraft	100%
Ilyushin Aviation	Feb. 2006	United Aircraft	51%
Gagarin Komsomolsk on Amur Aircraft	Feb. 2006	United Aircraft	25.5%
Sokol Aircraft	Feb. 2006	United Aircraft	38%
Chkalov Aircraft	Feb. 2006	United Aircraft	25.5%
Tupolev	Feb. 2006	United Aircraft	65.8%
OMZ Heavy Machinery			100%
Kamov Helicopter			100%
Transneft Pipeline			100%
Svyazinvest Telecom			75%
Rostelcom Telecom			38.1%
Aeroflot Airline			51%
United Energy Systems Electricity			52.7%
Alrosa Diamonds			32%
Rosoboronexport	proposed 2007	Rosteknologi	

does it. In the case of Yukos, the state and/or Putin reasserted control of Yukos by putting Khodorkovsky in prison and harassing over two dozen of his associates by either jailing them or threatening them with jail. In the meantime, the state picked up the pieces of Yukos at laughable fire sale prices. The state also employed crude tactics against Shell at Sakhalin II, BP in Kovykta, and Total in Kharyaga. Of course, almost every foreign operator in Russia is subjected to close, sometimes too close, supervision. Exxon-Mobil, for example, as of August 2007 had been subjected to ninety inspections at its Sakhalin work site. This is not to claim that the Western companies were completely innocent of the charges made against them or to deny that other countries often harass energy companies operating within their borders. But without an independent court of appeal to adjudicate these complaints and insist on due process, Gazprom or other state surrogates seem to feel no hesitation in launching campaigns of harassment that force the foreign companies involved to yield a controlling share to Gazprom for either nothing or a vastly underpriced sum.

Faced with a state determined to regain what it considers to be its priceless and historic legacy, the foreign partners were given no choice but to surrender. As Daniel Yergin of Cambridge Energy Research Associates has noted, this is not the first time energy resources around the world have been nationalized or for that matter in Russia itself.[84] This is Russia's ball game, not to mention their ball, and bat, and playing field, so they can do what they please. What is disappointing is that they are not doing it in what the Russians would call "a civilized way." Perhaps there is no "civilized way" acceptable to those who feel their property is being stolen, but if Russia wants to be—as indeed it feels it deserves to be—a member of the G-8 group of developed and democratic market economies, it will have to discipline itself from returning to the ways of its past. Instead, it should adopt less peremptory and more lawful methods of regaining control over its natural resources.

6

Natural Gas

Russia's New Secret Weapon

POWER OUT OF A PIPELINE

While its petroleum exports have generated the cash blizzard that has made Russia rich and allowed it to repay most of its foreign state debt, its natural gas and monopoly control of the gas pipelines that transport the gas to the West have transformed Russia from an anemic and essentially bankrupt charity case into a robust energy superpower with restored political muscle.

Initially it seemed like such a sensible idea. Determined to reduce their over-dependence on energy from the problematic Middle East, European leaders in the mid- 1980s, especially Helmut Kohl and later Gerhard Schroeder in Germany in 1998, concluded that Germany should diversify its sources of supply. One way to do this would be to support efforts to tap into energy exports from the USSR and its most important successor state, Russia.[1]

This required some rethinking by the major Seven Sister oil companies. Historically, they have worked to prevent the sale of Soviet discounted crude oil so that it would not undercut market prices in the capitalist world. This began to change in 1973. Following the lead of Eni (Ente Nazionale Idrocarburi) of Italy, which began to buy Soviet crude oil as early as 1931, Western petroleum companies began to view imports of petroleum from the USSR in a more positive light. This coincided with their need to diversify their sources of supply. The 1973 Arab petroleum embargo that accompanied the Yom Kippur War taught the West that manipulation of energy supplies not only could have important financial repercussions but could also be a powerful political tool.

In their effort to diversify, European leaders also decided to broaden energy use in Europe to reduce their overreliance on coal and petroleum. In France this took the form of an ambitious expansion of nuclear energy. By 2004, nuclear energy accounted for 78 percent of France's electricity. For environmental reasons, the Germans were more hesitant about nuclear energy, but for a time even they used nuclear energy to generate 30 percent of their electricity. But having decided to phase out nuclear energy, German leaders needed to find additional sources of power. Because of their physical proximity to the USSR, they agreed to supplement the natural gas they were beginning to use from the North Sea with natural gas delivered by pipeline from the USSR. Soviet gas would allow Germany to reduce its overreliance on petroleum, the risk of a nuclear accident, its exposure to turmoil in the Middle East, and the need to ship tankers through the Persian Gulf and other potentially dangerous open sea routes. In addition, manufacturing the pipe and the compressors needed to move the gas would generate jobs throughout Europe. The downside of the Soviet gas pipeline option was that it would put Germany at the mercy of a Cold War adversary. Most Germans still remembered the Berlin Blockade of 1948.

For those who had forgotten the Berlin Blockade or were too young to have experienced it, Ronald Reagan, when he subsequently became U.S. president, did all he could to remind them of how vulnerable they could become. Reagan understood the geopolitical risks that such a pipeline would create. He was very concerned that by building such a pipeline, Germany might some day find itself held hostage to Soviet demands. Given the German determination to diversify their sources and types of energy, however, the Germans regarded Reagan's arguments as unduly ideological and, even if it meant misleading Reagan as to their intentions, went ahead with the pipeline construction.[2]

Adding muscle to his rhetoric, Reagan banned the export of General Electric compressors and pumps, the preferred technology used in most of the world's gas pipelines. When the pipeline contractors sought out non-U.S. manufacturers, Reagan followed suit by ordering that similar bans would also apply to any non-U.S. manufacturers that utilized U.S. technology or parts in their products. This created a rift with his otherwise ideological soul mate, Prime Minister Margaret Thatcher of England.[3] Close as she was to Reagan, her first allegiance was to the British public. She wanted the jobs that would come from building the pipeline compressors that would go to the British company John Brown. It could easily fill in if GE could not. In the end, she ignored the U.S. demand that England impose export restrictions and instead allowed John Brown to build and export the necessary compressors.

The Europeans were not unaware of the risks that would come with relying on Russian gas. For that reason they agreed to seek, develop, and promote alternative sources of supply, particularly those from Norway in the North and the Barents Sea off Norway.[4] They also agreed that they would limit their use of Soviet gas to 30 percent of overall consumption, a promise they soon forgot. Not that he could do much about it, Reagan understood that with time all such initial caution would probably fade from memory and both European homeowners and industrial consumers would become more and more comfortable accepting gas imports from the USSR.

No matter what kinds of precautions are taken, a halt in the flow of natural gas exports that lasts more than a few days inevitably is disruptive. As a consequence, once manufacturers and households begin to incorporate imported natural gas into their daily work and living routines they are at the mercy of the exporter. That is almost certain to have political ramifications. Western leaders would have to think twice before resisting the political demands of the supplier.

Moreover, because natural gas pipelines, including the proposed Bratsvo ("Brotherhood") pipeline from the USSR that Reagan was trying to halt, are so expensive to build, it is simply not feasible to build a standby pipeline for emergency use. Since all but a small proportion of natural gas sold in the world comes via pipeline, should the flow through a part of one of those pipelines be disrupted—whether because of the weather, human mistakes, or political mischief—the pipeline-dependent consumer becomes particularly vulnerable.

One of the few possible alternatives to pipeline-delivered gas is LNG (liquified natural gas), but this, too, is very expensive and generally not a suitable standby for emergency use. Building the processing units needed to liquefy the gas at the exporting site and reconverting it at the importing site is very expensive. So are the specially built tankers that transport the gas. Building the combined LNG processing plant package often costs almost as much as building a pipeline. Consequently, such LNG systems are normally constructed only if the exporter and importer are willing to commit to long-term contracts similar to those signed by the parties utilizing a pipeline. That explains why, unlike the way petroleum is bought and sold on spot markets, there is still only limited use of a spot market for LNG. The result is that once a gas pipeline is built, it acts, as we said, like an umbilical cord. Severing it is bound to be disruptive.

PUTIN REINS IN GAZPROM

It was good luck that his rise to power coincided with a tightening in world energy markets; but in retrospect it seems clear that Putin understood as early as 1997 that with its oil and gas reserves and pipelines, Russia was well situated to take advantage of this new dynamic.[5] While Saudi Arabia has the world's largest reserves of crude oil, Russia, not Saudi Arabia, has the world's largest reserves of natural gas. Most experts agree that Russia holds 27–28 percent of the world's natural gas reserves.[6] With a little more than half of what Russia has, Iran with 15 percent of the world's reserves ranks second in size of natural gas reserves. Qatar is close with 14 percent. Even though Canada is a major supplier of natural gas to the United States, it has only 1 percent of the world's reserves. Since no other country but the United States, Saudi Arabia, and the United Arab Emirates has even as much as 3 percent of the world's natural gas, Russia is in a dominant position.[7] Its reserves and its pipelines, if strategically utilized, have the potential to provide Russia with a powerful political and economic weapon. To his credit, Putin understood this potential and has been skillful in utilizing it.

Building on his concept of "national champions," as we saw, Putin's first priority was to purge the self-dealers and asset-strippers from Gazprom. He seems instinctively to have recognized that Gazprom would make an ideal flagship, on the assumption, of course, that he could find managers who would place the interests of the state above their own. That is why almost immediately after his election as president, Putin sought to put managers in place who would no longer strip off producing assets into their privately held empires. To take on the task, Putin began to appoint comrades he considered loyal and trustworthy, almost all of whom were FOP (Friends of Putin) from St. Petersburg. He knew them from his days either in the KGB or in the mayor's office when Putin headed the office of International Affairs under Mayor Anatoly Sobchak.

Not everyone in Moscow was happy with these "provincials" from Russia's second city taking charge of what had been the political center for eighty years. We saw in Chapter 5 how Putin began by removing Chernomyrdin in June 2000 from his post as chairman of the Gazprom board. He replaced him with Dmitri Medvedev, who also took on the job as head of the Kremlin administration. (In 2007 Putin chose Medvedev again, this time as his successor for president of Russia.) Medvedev had previously worked in the St. Petersburg mayor's office alongside Putin. The following year Putin replaced Rem Vyakhirev as Gazprom CEO with Alexei Miller, who had also worked for the St. Petersburg mayor.

From the outside, the transition within Gazprom appeared to be rather straightforward and routine. Both Chernomyrdin and Vyakhirev left without too much fuss. Chernomyrdin went on to become Russia's ambassador to Ukraine. But while it may have seemed routine, it was anything but. With all their spoils and patronage to protect, Vyakhirev in particular had fought ferociously against previous attempts to oust him. Those opposed to his tenure had much to criticize. Among other charges, some members of Gazprom's board of directors complained that the company had paid little in either taxes or dividends. In 1995 and 1996, despite having generated earnings of almost $2 billion, Gazprom paid only $3.5 million in dividends to the state.[8] Even stranger, the state at the time held 38.4 percent of the company's stock.

Stingy as they were with dividends and taxes, the managers were overly extravagant in using company funds to pay bonuses to themselves and build resorts for the exclusive use of the staff. Others complained about what they considered the waste of money spent in building the company's Taj Mahal–like corporate headquarters[9] (see Introduction). This was all in addition to the asset stripping.

There were also suspicions that the company's accounting statements did not reflect the true financial situation. In 1999, for example, based on Russian accounting standards, Gazprom reported a profit of $1.3 billion. However, when calculated according to Western accounting practices, Gazprom had a loss of $3.2 billion.[10]

Putin ultimately succeeded in changing Gazprom's senior management, but others had tried earlier and failed. The difficulty is illustrated by what happened when Boris Fedorov tried to convince his fellow members of the board of directors to join him in bringing about a change in management. Before he became a member of Gazprom's board of directors, Fedorov had been Russia's Minister of Finance and for a time the director of the Russian Tax Office. Now as an investor in Gazprom, he sought to clean up the company. One of his goals was to bring in a new auditor. He wanted such an auditor to come up with a "second opinion" on the relationship between ITERA, that Florida-based company that at one point was Russia's second largest producer of natural gas, and Gazprom. Most important, Fedorov began to call openly for Vyakhirev's immediate removal as CEO before his term expired.

Vyakhirev did not take kindly to Fedorov's effort. Fedorov told me that he began to fear for his life, particularly after he was visited by representatives of the Russian mafia. Then as if it were all a scene from the movie *The Godfather*, someone poisoned Fedorov's dog![11] If there were any doubts as to what was happening, more than fifty newspaper articles in the Moscow press suddenly and simultaneously appeared with vicious attacks on him. Only when Vyakhirev was fired by Putin in July 2001 did the attacks abruptly came to an end. Intrigued by what seemed to be the obvious orchestration of this effort, Fedorov subsequently canvassed each newspaper to see what had precipitated this sudden campaign. As if they were normal events, each paper explained that such attacks were a common occurrence, a normal part of the for-hire nature of Russian journalism. He managed to compile a price list indicating how much each paper charged for these attacks. Reflecting the market, the higher quality newspapers such as *Vedomosti*, which is jointly owned by the parent companies of both the *Wall Street Journal* and the *Financial Times*, charged the highest rate: $6,000 for each of the four articles they published.

No one knows what might have happened if Vyakhirev had been able to continue his campaign against Fedorov. Fortunately for Fedorov, he and Putin had similar agendas. Putin was just as eager as Fedorov to put an end to the asset stripping, the self-indulgent extravagance, and failure to compensate the state and other stockholders for their investment. But unlike Fedorov, Putin had the power to implement it. Yet Putin also had a supplemental and—in his view—equally important agenda. He wanted Gazprom to become the first of what he had hoped would be those "national champions."

Once in charge, one of Medvedev's and Miller's first assignments was not only to bring a halt to any further asset stripping but to reclaim assets that had been stripped earlier. This was not easy to do, but Miller moved aggressively. One of his first targets was ITERA. From the mid-1990s, ITERA had operated as a middleman between a bunch of countries such as Ukraine, some of the Caucasus countries, and Central Asian producers. In the process, ITERA earned a handsome profit for that Florida-based corporation whose trustees were mostly associated in one way or another with Gazprom management. To persuade it to cooperate, Miller denied it access to the Gazprom pipeline, a step that by 2004 all but forced ITERA into bankruptcy.[12] In 2006, faced with an offer they could no longer refuse, ITERA's managers agreed to resell their 51 percent interest in Sibneftegaz back to Gazprombank for the bargain price of $130 million.[13]

Despite Miller's success with ITERA, Gazprom would not become a transparent corporation overnight. As we shall see shortly, Gazprom's dealings and interactions with the state and other companies remained almost as opaque as before.

GAZPROM, THE HOLY OF HOLIES

Repeatedly, Putin has signaled how central Gazprom is to him and the role it must play in Russia's emergence as an energy superpower. He has referred to it elsewhere as this "holy of holies."[14] Given Gazprom's role in his thinking, it is not surprising that in his May 2006 state of the nation speech, Putin took time to boast that Gazprom had just become the world's third largest corporation as measured by the total value of its stock. At the time, only Exxon-Mobil and General Electric were larger. (Microsoft subsequently increased in value to push Gazprom to fourth place, and it in turn and even Exxon-Mobil were displaced by a set of Chinese corporations when an index of Chinese stocks more than doubled in 2007–2008.) Admittedly, such information would be of interest primarily to readers of business newspapers, but it is unlikely that many other world leaders (except those fighting an inferiority complex) would choose to emphasize such a fact in their state of the nation speech. By including it in his presentation, Putin signaled its importance to him. Yet Putin and those around him have even higher ambitions. Putin's ultimate goal is to see Gazprom overtake and surpass (Nikita Khrushchev's favorite way of comparing the USSR and the USA) Exxon-Mobil to become the corporation with the largest capitalized value in the world. Moreover, as he has put it, he sees no reason why some day the value of Gazprom's stock should not rise from $300 billion to $1 trillion, overtaking Exxon-Mobil along the way.

In Putin's mind, Gazprom's emergence as a dominant international corporate player was no accident. In that same 2006 state of the nation speech, he went so far as to claim this was "the result of a carefully planned action by the state."[15] While patting himself on the back for bringing to life this national champion, Putin somehow ignored that perhaps the post-1998 increase in world energy prices might have had a little to do with that surge in the price of Gazprom stock. Putin is not the only Russian official who associates the turnaround in Russia's fortunes with whether Gazprom is thriving. As we noted in the Introduction, Alexander Medvedev, deputy CEO of Gazprom and general director of Gazexport, its export affiliate, frequently seeks to reassure foreign audiences by insisting that what is good for "a strong Gazprom is good for the world."[16]

Given the symbiosis between Gazprom and Russia, Putin and his colleagues do not take kindly to those who question how Gazprom is run. William Browder, the grandson of the long-time head of the U.S. Communist Party, Earl Browder, is a good example. As the founder and director of the $4 billion Hermitage Capital Management investment fund, Browder the younger has been an outspoken advocate of investing in Russia. However, Browder has criticized not only the original Gazprom executives but Putin's subsequent appointees. Browder concedes that such Putin initiatives enhance the glory of Russia, but they are not in the best interests of the company's stockholders and a higher return on their investment. Browder soon discovered that while it may be okay for Putin to criticize Gazprom's previous management, Putin and his subordinates in the Kremlin are not eager to have others, especially foreigners, do the same. To register its displeasure with Browder and alert others that there are limits to criticism of this "holy of holies," the Russian government canceled Browder's Russian visa and prevented him from returning to his home in Moscow when he left Russia in 2006.

As the Browder incident illustrates, it is hard to tell where Putin begins and Gazprom ends. Alexander Medvedev, deputy chairman of Gazprom's management committee, insisted in a presentation in St. Petersburg on June 21, 2007, that to the contrary, Gazprom operates free of interference from the Kremlin. As he put it, "We don't get hourly calls from the Kremlin. We get none at all." That would seem to overlook not only Putin's assertion about his successes with Gazprom and some of the other national champions but also Putin's actions and unwavering advocacy and support for Gazprom's initiatives. As the wags have it, Russia or "Gazpromistan" is run by its president and spiritual leader, Gazputin, an obvious play on the gas-rich countries of Central Asia, as well as Rasputin, the mad monk favorite of the last czar's wife, Czarina Alexandra.[17]

Putin's new appointees, Chairman Dmitri Medvedev and particularly Alexei Miller, moved quickly to stop the asset stripping. It was not easy to repair all the damage, but as a minimum, they put an end to further dismantlement.

The next step was more controversial and most likely another example of a Putin-initiated action. For many years, Gazprom, like the Ministry of Gas before it, intentionally held down the price of natural gas it sold within the boundaries of the former USSR, far below comparable market prices in the West. This was done to facilitate Soviet industrialization. But it also had the effect of encouraging the wasteful overuse of all raw materials, particularly oil and gas. Because the price was so cheap and because there always seemed to be more oil and gas available, there was no need to worry about conservation. This policy continued for several years after the breakup of the USSR. Prices were kept below world prices not only within Russia but also in the other republics that made up the USSR.

UKRAINE IS TOLD TO PAY THE MARKET PRICE

It was easy to ignore these hidden subsidies that came from being a part of the USSR, but they were substantial. Shortly after assuming the role of president of Ukraine in January 2005, Victor Yushchenko adopted a noticeably cooler attitude toward Russia. At the same time, he drew closer to the West, including the United States. In reaction, pushed by Putin, Gazprom began to warn that a looser alliance would lead to an end to gas export subsidies. If Yushchenko wanted a closer relationship with the West, he should also be prepared to pay prices closer to those paid by Western customers. As Putin told a group of us in September 2004, Yushchenko was welcome to seek a closer alliance with the West and turn his back on Russia, but he should understand that if he did so, Russia was under no obligation to continue to subsidize its energy exports to Ukraine. Ukraine was paying as little as $50 per 1,000 cubic meters while the market price in the West at the time was $150 per 1,000 cubic meters, so paying the higher price would cost Ukraine $3–5 billion a year. Since the United States was providing Ukraine only about $150 million in aid at the time, turning its back on Russia would be costly. So in Putin's words, "Ukraine should think twice about any such embrace of the West." By contrast, Belarus, which was then a close Russian ally, was charged less than $50 per 1,000 cubic meters for its deliveries, not much different from what users within Russia itself had to pay in 2006.

Warning that it was prepared to take extreme measures, on January 1, 2006, Gazprom demanded that Ukraine pay $150 per 1,000 cubic meters, a threefold increase from the earlier charge. Refusing to be intimidated, Ukraine insisted on paying the lower fee, arguing that this lower price had been agreed to during previous contracts. Any reduction or cessation of gas deliveries through the pipeline by Gazprom would be a contract violation. In response, Gazprom insisted that the contract had expired and proceeded to reduce the flow of gas, sending through just enough to meet its contract obligations to its customers in Western Europe. Ukraine, however, continued to withdraw the same amount of gas from the pipeline that it had prior to December 31, 2005. Like Belarus, it felt entitled to pay for those deliveries at the lower price. The Russians then reduced the flow to Ukraine. Claiming Gazprom had broken its contract, Ukraine provided first for its own needs and only then sent what gas was left on through the pipeline to the West.

Gazprom and Putin, however, used an economist's arguments, pointing out that Russia was only asking Ukraine to adhere to market practices and prices. In the long run, this would be good for Ukraine. Gazprom was only helping Ukraine wean itself away from distorting subsidies. Isn't that what the United States and the West Europeans had been urging Russia to do? Accordingly, when the flow of gas was reduced, Gazprom spokesmen repeatedly insisted that none of this pressure on Ukraine was political. The flow of gas would be resumed once the Ukrainians agreed to pay the market price, with the emphasis on market price. To the contrary, it was not Russia that was at fault but Ukraine. By diverting the gas intended for Western Europe to itself, the Ukrainians were simply stealing Europe's gas.

THE USE OF INTERMEDIARIES — WHO IS THE REAL OWNER?

To Russia's surprise, however, most Europeans turned out to be more sympathetic to Ukraine than to Russia. It was January, after all, and cold, and turning off the gas was not a nice thing to do. Moreover, there was widespread suspicion that executives of the gas companies in both Ukraine and Russia were using the crisis to stuff their own pockets with kickbacks. These suspicions arose because the final agreement did not involve a direct and transparent contract between GazpromExport (Gazprom's export division) and its Ukrainian counterpart. Instead, GazpromExport agreed to deliver gas from Russia and Turkmenistan to a mysterious company called RosUkrEnergo (RUE), which in turn sold it to UkrGaz-Energo, the Ukrainian utility that delivers it to Naftogaz, which ultimately delivers the gas to the actual Ukrainian consumers.[18] Setting up these different entities was designed to confuse outsiders and, as we shall see, those in charge succeeded brilliantly.

RUE first began to supply UkrGaz-Energo in 2005. The use of intermediaries, however, dates back to the 1990s when ITERA, that opaque Florida-based company, stepped in to deliver gas from Turkmenistan to Ukraine. ITERA was created by Igor Makarov. A native of Turkmenistan, Makarov was a poor boy who became a world-class bicycle racer, bringing glory to Turkmenistan and becoming a local hero. As a result he was befriended by the Turkmen president, the President for Life or Turkmenbashi, as he called himself, Saparmurat Niyazov. (Except that he was president of Turkmenistan and not Kazakhstan, Niyazov could have served as a model for the movie *Borat*.) After the collapse of the Soviet Union, Turkmenistan found itself with almost no convertible currency and as a result, in serious need of basic consumer goods. Through his friendship with Niyazov, Makarov was given access to Turkmenistan's natural gas, which Makarov was then allowed to use to barter for food and other consumer goods. To do all of this, of course, Makarov also had to convince Rem Vyakhirev of Gazprom to grant him access to Gazprom's pipeline network. Gazprom was not interested in bringing gas into Russia to compete in its own domestic market but it was willing to transport Turkmen gas to Ukraine, which it did beginning in 1994. This earned ITERA a handsome profit (and who knows what for Vyakhirev), at least until the financial collapse of August 1998. The market collapse, while bad for most businesses, including Gazprom, provided ITERA with a great opportunity to buy up some distressed Gazprom properties at an auction. Since auctions in Russia are not noted for their transparency, especially those associated with the Loans for Shares privatization of 1995–1996, it is hard to dispel the suspicion that with Vyakhirev's blessings, ITERA ended up with valuable Gazprom assets at a cost that was low even compared with the distressed prices that prevailed in 1998. In any case, after Putin removed Vyakhirev as CEO of Gazprom, Alexei Miller, Vyakhirev's successor, moved rapidly to repossess as much as two-thirds of the property that ITERA had acquired from Gazprom.

It may have been coincidence, but once Vyakhirev was no longer in charge of Gazprom, ITERA lost one of its main protectors. Not only did ITERA have to return some of its assets to Gazprom but it was also squeezed out of the Turkmenistan-Ukraine trade in 2002 by a company controlled by Dmytro Firtash called Eural Trans Gas.

Eural Trans Gas in turn held on to the franchise until 2005 when RUE took it over.

All of these companies were opaque and regarded with suspicion.

Ostensibly, while the companies were different, it seems that many of the principal owners of the various companies remained the same.[19] Thus, Dmytro Firtash, a Ukrainian businessman who had been bartering consumer goods to buyers in Ukraine, Turkmenistan, and Russia, joined a company called Highrock Holding in 2001. Firtash insists that one of his partners in Highrock Holding was Igor Makarov, president of ITERA. Subsequently, ITERA was superseded in Turkmenistan-Ukraine trade in 2002 by Firtash's Eural Trans Gas. According to the *Financial Times*, Makarov subsequently denied he ever had an economic interest in Highrock Holding.[20]

That is of interest because it reflects the murkiness associated with importing natural gas and the unsavoriness of the parties involved. When Putin moved to clean up Gazprom, the new Gazprom management in turn began to apply pressure on ITERA, not only to return assets to Gazprom (as we saw) but to force it out of the transit business between Turkmenistan and Ukraine. That opened the way for Eural Trans Gas, which hardly seemed much of an improvement.[21]

The U.S. Department of Justice and the FBI have been investigating whether Russian and Ukrainian mafia members have also been involved with these companies. After it became known that the FBI was investigating Highrock, its principal owner Dmytro Firtash acknowledged that he and his junior partner, Ivan Fursin, owned Centragas Holding, which owns 50 percent of RUE.[22] The other 50 percent was owned by Arosgas Holding, Gazprom's Austrian affiliate. Without meaning to deprecate Ukrainian and Austrian skills at obfuscation, this is almost but not quite as complicated as trying to ascertain who owned what of the Enron Company in Texas before it went bankrupt. Gazprom demanded part ownership in RUE for the obvious reason that its monopoly control of the gas pipelines in Russia gave it the exclusive rights to ship the gas from Turkmenistan and its neighbors through Russia to Ukraine.[23]

Firtash's role was news because for some time his involvement had been shielded from public view by Raiffeisen Investment AG, the investment arm of the Raiffeisen Bank of Austria. Raiffeisen Investment AG acted as trustee for what was thought to be the true owners. It should be noted that the Raiffeisen Bank itself has had a long history of dealing in Eastern Europe and the Soviet Union, which almost guarantees that many of its transactions will not be transparent.

For Gazprom, a Russian company, to be involved in such convoluted corporate juggling within Ukraine is troubling enough, but what drew the attention of the FBI was the possible evidence that another "Ukrainian businessman," Semion Mogilevich—or at least his wife— had also been involved in these machinations as a partner with Firtash in Highrock Holdings.[24] Mogilevich has, as they say, been a person of interest to the FBI since 2003. Although he was found not guilty of criminal activity in one trial, he has been on the FBI's Most Wanted List, regarded as "one of the world's most sophisticated international criminals."[25] In January 2008, after considerable international pressure, Russian authorities finally arrested him. Among other crimes, the FBI wanted to question him about his alleged involvement in prostitution, drug trafficking, and stock fraud.[26]

The shenanigans of Highrock, RUE, ITERA, Eural Trans Gas, and Gazprom illustrate the skill with which veterans of the black market in the Soviet era have learned to manipulate the market system and obfuscate their operations from even sophisticated investors. How they learned to build such a web of false fronts and hidden assets for themselves despite having grown up in a system of relatively simple central planning remains a mystery. Where did they learn such sophisticated schemes? It was not taught to them in Soviet institutions of higher learning or in Gosplan, the state central planning agency.

Such shadowy entities linking up Russia and Ukraine cast doubt on the integrity and the transparency of the economic interactions between Russia and Ukraine. The Russians' argument that Ukraine deserved to pay the market price for gas was weakened when it became known that at least 60 percent of the gas supplied to Ukraine actually came from Turkmenistan, not Russia. Nor did Gazprom win sympathy for itself when it was learned that at the time Gazprom refused to pay Turkmenistan more than $46 per 1,000 cubic meters while selling the same gas to RUE and Ukraine for $95 per 1,000 cubic meters. Only in February 2006 did Gazprom agree to pay a comparable amount for its Turkmenistan purchases. Gazprom control of the pipeline linking Turkmenistan with the West, a legacy of the Soviet era, allowed Gazprom to squeeze Turkmenistan this way. Gazprom agreed to raise the price to $130 in 2008, but that remained for below the $354 Gazprom expected to collect from its sales to Europe.

What does seem odd is that Turkmenistan agreed to extend a PSA (production sharing agreement) to Russian companies engaged in energy development there. Russia has been invalidating similar PSAs it made earlier with Western companies such as Shell, Exxon-Mobil, and Total. While Russian companies have opposed extending PSAs to foreign companies working within Russia, evidently that has not prevented Russian companies from signing up for similar concessions for themselves in other supplicant states.

What all this illustrates is that Gazprom control over the pipeline network of the one-time USSR Ministry of Gas has been one of Russia's most valuable and strategic assets. For the time being, the only way Central Asian countries can export their gas to Europe is through Gazprom's pipeline.

SEEKING A WAY AROUND GAZPROM PIPELINES

Efforts are afoot by the United States and some members of the European Union to build an alternate gas pipeline under the Caspian Sea from Central Asia to Baku. U.S. Vice President Dick Cheney made a visit to Kazakhstan in the spring of 2006 to seek support for such a bypass. That gas pipeline would then parallel the recently completed private petroleum pipeline from Baku, Azerbaijan, through Tbilisi, Georgia, which then terminates in Ceyhan, Turkey, on the Mediterranean Sea. There is some uncertainty, however, as to whether such a gas pipeline will be financially viable. To ensure that it would not be, Putin moved immediately to neutralize Cheney's effort by making a follow-up visit to dissuade Kazakhstan from such a move.

As we saw in the Introduction, the Russians, along with the Italian company, Eni, are doing everything they can to ensure that a bypass and diversionary pipeline is not built. Putin is doing this by trying to dissuade not only potential suppliers but customers from using such a pipeline. Toward that end, Russia and Putin have reached at least a tentative agreement with Turkmenistan to tie up much of its natural gas exports for twenty-five years so little would be available for an alternative routing.[27] Unless the U.S. government and other promoters of such a pipeline can assure themselves and prospective users and investors that the volume of gas carried by the pipeline will be large enough, they will not put up the money needed to build it.

As for Georgia, given its crucial role as the connecting link between Baku and Ceyhan, Russia has done its best to destabilize the region and keep Georgia from operating the pipeline in an orderly and reliable way. If Georgia collapses in turmoil, investors will not put up the money for a bypass pipeline and Russia will be able to maintain its pipeline monopoly. That, at least in part, explains why the Russian government has provided rather open support for South Ossetia and Abkhazia, two regions that seek to separate from Georgia and align themselves instead with Russia.

That is not all Russia has done. In 2006, after Georgia arrested and then expelled some Russian embassy officials on charges of espionage, Russia declared an embargo on imports of Georgian wine and mineral water as well as its fruits and vegetables—its most important export earners. At almost the same time, Russia shut down transport and postal service to Georgia, thereby severing its most important link to the outside world.[28] To underline its hostility, Russia also expelled Georgians living and trading in Russia, not only those without legal documentation but also those who had proper permission. In addition, there were disruptions in the flow of electricity from Russia. At about the same time in January 2006, the gas pipeline passing through North Ossetia from Russia to Georgia mysteriously exploded.[29] This coincided with the campaign against Ukraine and the application of similar pressure on Moldova. As with Ukraine, the Russians demanded that Georgia and Moldova agree to pay the much higher Western European market price for gas.

It has not been an easy time for either Georgia or Moldova. In the fall of 2007, for example, opposition groups in Georgia began to call for the resignation of Mikhail Saakashvili's pro-Western government. Saakashvili called out the troops and put down the demonstration in a rather heavy-handed way, insisting that these protests were provocations organized by Moscow in an effort to regain control of the area, an accusation Moscow disputes.

Upping the ante, Gazprom began to demand that it be given ownership of Georgia's and Moldova's domestic pipelines. In 2007, the two agreed to pay more for gas—in Georgia's case, a price of $235 per 1,000 cubic meters, about the same as Europe—but Georgia refused to yield to Gazprom demands that it sell off its domestic pipelines. But Moldova, along with Armenia, both succumbed and agreed to sell Gazprom a controlling share in their gas distribution networks.[30]

Undoubtedly such pressure on relatively small countries hurts. But despite the harassment, the Georgian economy, in particular, has enjoyed an unprecedented economic boom. It may have been an effort to make Russia and Putin look foolish, but President Mikhail Saakashvili attributed the country's astounding 10 percent annual growth to the embargo itself. As he explained, the embargo unexpectedly led to an increase in foreign direct investment from Kazakhstan and the United Arab Emirates.[31] (By investing this way Kazakhstan seemed to be doing what it could to undermine Russia's policy. However in May 2007, after a series of visits by Putin, Kazakhstan appeared to become more supportive of Russia.)[32] It also helped that President Saakashvili's government began to implement a vigorous program of economic reform that included the adoption of a 12 percent flat tax (1 percent lower than a similar flat tax in Russia), the slashing of red tape, and the introduction of a new customs code. Moreover, the cutoff of gas from Russia caused only temporary hardship. Almost immediately the Georgians were able to arrange for substitute deliveries, primarily from Azerbaijan. By 2007, Georgia had managed to shift more than 80 percent of its natural gas imports to non-Russian sources.[33]

While Russian intimidation of Georgia may have backfired, the Russians continued to harass Ukraine, and in 2007, even their heretofore cooperative ally, Belarus. Much to the disbelief of Alexander Lukashenko, president of Belarus, in January 2007 Russia began to apply the same type of pressure on Belarus that it had on Ukraine a year earlier. This was quite a surprise. Lukashenko, a one-time collective farm manager and what some have called the last dictator in Europe, has used his powers to tie Belarus almost blindly to Russia. It was a shock therefore when Belarus was told it too would have to pay more for its gas. At first the Russians demanded $200 per 1,000 cubic meters. Ultimately, they consented to a price of $100 per 1,000 cubic meters. But even $100 meant a doubling of prices from 2006. Belarus was also asked to pay $180 for each ton of petroleum sold to Belarus as a form of export duty.[34] Reluctantly, Lukashenko agreed to the $100 per 1,000 cubic meter price for gas. But to Lukashenko, this was much more than an unfriendly gesture from what he had frequently boasted was a supportive partner. Belarus depended on these highly subsidized and therefore cheap petroleum imports from Russia and their subsequent re-export to Western Europe at considerably higher world prices. The difference between what it paid and what it charged provided Belarus with a substiantial profit, which is used to underpin its otherwise shaky economy.

In response to the imposition of the $180 Russian export duty, Belarus then imposed a $45 a ton transit fee on the petroleum the Russians were sending on to Western Europe. Since about half of all Russian petroleum sold to Western Europe is shipped through Belarus, this was a significant countermeasure.[35] After almost a week and a half of nasty words and an occasional halt in the flow of gas and oil to Belarus, the two sides reached a compromise. Russia agreed to lower its export duty on petroleum from $180 a ton to $53, and, in turn, Belarus agreed to abolish its transit fee.[36] But even though Lukashenko agreed to the higher prices, Belarus fell behind in its payments so that by mid-2007, it was $456 million in arrears. Once again Russia threatened to cut off deliveries. After apparently turning for help to Hugo Chavez, president of Venezuela, Belarus paid its bill and the flow of gas was no longer interrupted.[37]

THE PIPELINE POKER GAME

While most of the outside world's attention was on whether Belarus, like Ukraine the year before, would agree to pay a higher price, at the same time there was an even more significant effort by Russia to expand its control of the pipeline network linking Russia to its European consumers. After the collapse of the USSR, almost all of Gazprom's customers in the CIS, the Commonwealth of Independent States (the former Soviet republics), found themselves with significant bills they could not pay for the gas they had already imported. In what became a standardized routine, Gazprom would then offer to cancel the debt or charge a lower price if the Ukrainians, Armenians, Moldavans, or Georgians would give Gazprom an equity stake in their domestic pipeline network (see Table 6.1). In Belarus, Gazprom offered $2.5 billion for a 50 percent stake in Beltransgaz, which owned the gas export pipeline.[38] While Belarus agreed, some of the others have held back. The *Financial Times* for example, reported that in Ukraine, President Viktor Yushchenko had publicly criticized Naftogaz for offering Russia a measure of control over its gas transit system. Yushchenko and his allies feared that if Gazprom gained an equity interest in their pipelines, Russia would demand an ever larger say in their economic and political affairs.[39]

There was particular concern that if Russia or Gazprom were allowed to buy up local gas distribution systems used by its customers to maintain their monopoly control and economic rent, the Russian operators would do all they could to exclude other potential suppliers. In fact, gaining control over pipeline access to other producers of gas as well as to Gazprom customers has been a major goal of Gazprom and Russia. It is not only the foreign consumers of Russian natural gas who worry that Russia will some day control gas pipelines within their territory; non-Russian producers of natural gas operating in what used to be the USSR are also very much concerned. They are indeed vulnerable. As long as there is no other way for the Central Asian countries—or for that matter Russian petroleum companies—to transport their gas to Europe except through Gazprom-controlled pipe-

TABLE 6.1   Gazprom Expansion Abroad

	Share of Domestic Pipeline	Share of Transit Pipeline	Direct Sales to Consumers	Campaign to Gain Pipeline Contract
Austria			×	
Belarus	×	×		×
Bulgaria	×		×	
Estonia			×	
France			×	
Georgia				×
Germany	×	×	×	
Greece				×
Hungary		×	×	
Italy			×	
Latvia			×	
Lithuania			×	
Moldova				×
Poland		×		
Portugal	✓			
Serbia	×	×		
Turkey	×			
Ukraine			×	✓
United Kingdom			×	

Key: × already acquired   ✓ negotiating
Source: Financial Times, December 21, 2006, p. 4; Wall Street Journal, January 28, 2008, p. A16.

lines, the only alternative for the Central Asians is to find customers in Asia or accept a Gazprom-dictated price for their gas. That explains why until 2006 Turkmenistan was forced to sell its gas to Russia for as little as $46 per 1,000 cubic meters. Ironically, at that price, when the Russians were the sellers and Ukraine, Belarus, Georgia, and Moldova were the buyers, Russia complained that the price was too low. It also explains why landlocked Turkmenistan, as well as its neighbors Kazakhstan and Uzbekistan, express interest now and then in a proposal to build a pipeline under the Caspian Sea from Kazakhstan to Azerbaijan and from there overland through Georgia and Turkey and on to Europe. This was the proposal promoted by U.S. Vice President Dick Cheney when he went to Kazakhstan in the summer of 2006.[40]

For the time being, Nursultan Nazarbayev, the Kazakhstani president, has indicated he would keep his options open.[41] Most of the petroleum from Kazakhstan continues to move west overland to Russia through the pipeline of the Caspian Pipeline Consortium (of which the Russian pipeline monopoly Transneft owns 24 percent) to Novorossiysk on the Black Sea. But Kazakh petroleum is also flowing through the Baku-Tbilisi-Ceyhan route, thereby bypassing Russian territory and also the dangerous and overcrowded Bosporus Straits (see Figure 2, page 8).[42]

PREVENT A SOUTHERN ROUTE

To Putin, all of this is like a giant chess match. Every move by a rival must be met by Putin with an even more attractive offer. He was confronted with such a challenge in December 2006 when a consortium led by BP began construction of a South Caucasus pipeline designed to transport natural gas from the Shah Deniz field in the Caspian Sea, as well as gas from Turkmenistan, Uzbekistan, and Kazakhstan, through Azerbaijan and Georgia to Ceyhan, the Turkish port on the Mediterranean Sea. From there, the gas would be shipped to the Balkans and ultimately to the European Union.[43] The route would parallel the already built Baku-Tbilisi-Ceyhan petroleum pipeline. Because it passes through Georgia, this gas pipeline would also make gas available for Georgia. But the pipeline's main purpose would be to free countries in Europe from being so dependent on Gazprom. From Turkey, the gas would be shipped through the NABUCCO pipeline, which is scheduled to be finished by 2011. NABUCCO would carry gas through Bulgaria, Romania, Hungary, and Austria, and from there to the West.[44] The main transit and storage hub would be built in Austria by OMV, the lead promoter of the project.

Turkey has also proposed to work with Iran to ship its gas overland by pipeline through Turkey and on to Europe. If it is eventually built, such a pipeline might also be used to transit gas from Turkmenistan. Some gas from Turkmenistan is already shipped to northern Iran, for now the only outlet for Turkmen gas that does not flow through Russia.[45]

The big challenge for BP, the NABUCCO partners, and Turkey is to see whether they can sign up enough customers to make the effort profitable. Eager to see that they did not, Gazprom moved simultaneously to increase deliveries of gas to Turkey via its Blue Stream pipeline under the Black Sea. Blue Stream was officially inaugurated in November 2005. From Turkey the Russian gas will be piped on to Western Europe via Gazprom's South European Gas Pipeline (SEGP). A Gazprom delegation, headed as usual by Putin, went to Budapest in 2006 in an effort to convince the Hungarians that using Gazprom gas, not BP's from the Caspian, was a better deal for Hungary. Since the market would probably not be large enough to support both pipelines, Gazprom and Putin hoped in this way to preempt the NABUCCO effort.

To make the offer more appealing than the NABUCCO route, Gazprom proposed to offer its gas sooner and cheaper. It also sought to persuade Hungary, an essential NABUCCO partner, that it should support Gazprom's South European Gas Pipeline for Russian gas (SEGP) instead. To make it worth their while, Gazprom offered to provide Hungary with an attractive long-term supply contract, and to make the offer even harder to resist, Gazprom promised that under its proposal, Hungary rather than Austria would become the European hub.

Political grandmaster that he is, Putin's tactics appeared, at least initially, to have worked. On March 12, 2007, the Hungarian prime minister Ferenc Gyurcsany announced that Hungary would support Gazprom's Blue Stream pipeline rather than NABUCCO.[46] As he put it, NABUCCO is "a long dream and old plan. But we don't need dreams, we need gas." No doubt the fact that Hungary, not Austria, would be the hub for the Blue Stream project was also a factor. The Hungarian prime minister explained that because the European Union had yet to agree on a common energy policy, it was dangerous for Hungary to wait when it had the option of making a favorable bilateral deal with Gazprom and in so doing solve its immediate energy problems.[47] Gazprom already had available the gas that Hungary needed. The NABUCCO pipeline, however, had yet to be proven. At best it would not be available until some time in the future. Eager to support the NABUCCO alternative, the European Union argued otherwise. Azerbaijan already had enough gas available and Kazakhstan was in the process of adding more.

After Gyurcsany announced that Hungary had opted for the Russian Blue Stream variant, which ran through the Black Sea from Russia to Turkey, he and his government evidently had some second thoughts. Perhaps, he suggested, they should keep their options open at least a little longer. Showing that the Hungarians are good chess players as well, a spokesman for the Hungarian government told a press conference the next day that Hungary was still willing to work with the NABUCCO consortium. After all, NABUCCO was strongly supported by the European Union and, as a new member, Hungary could not disregard its wishes, at least in the early planning stages and especially since there were still uncertainties in both projects.[48] As Prime Minister Gyurcsany explained after meeting with President Putin, "Why shouldn't we receive half from one source and [half] from the other?"[49] The prospects for NABUCCO also improved when Germany's second-largest gas company, RWE, decided to join in as a sponsor.

Putin clearly seemed determined to prevent the construction of the NABUCCO pipeline. To add to the competition from the Blue Stream and SEGP gas pipelines, in the summer of 2007 Putin along with the Italian company Eni proposed the construction of what they called the South Stream pipeline. This would be yet another gas pipeline from Russia running under the Black Sea to Bulgaria and then on to Italy.[50] As if all these proposals and negotiations were not complicated enough, in July 2007 OMV, the partly state-owned Austrian energy company, tried to take over MOL, the recently privatized Hungarian national energy company. While OMV earlier had initiated the NABUCCO project, as NABUCCO appeared to flounder, OMV reversed course and agreed instead to a deal with Gazprom that would make Vienna a gas hub. In effect this would mean that Gazprom would be dropping MOL as its main partner and eliminating Hungary as the hub. The Hungarians were opposed to OMV for other reasons as well. In 2006 they had gone to the trouble of fully privatizing MOL. If the partially state-owned OMV were allowed to buy it up, the Hungarian MOL would again become a state company, only this time the state would be Austria. There was also concern that if Gazprom should some day acquire OMV then Gazprom would own gas distribution facilities not only in Austria but also in Hungary.[51]

BUILDING A BALTIC SEA BYPASS

The Caspian Sea pipelines via Turkey and the pipelines in the south of Eastern Europe are not the only instances of rivalry between Gazprom and Western companies and governments or where Putin has taken it upon himself to represent Gazprom. Even though Ukraine, Belarus, and Poland have complained that they were victimized by Gazprom, one of Gazprom's biggest concerns has been to find a way to protect itself from Ukraine and Belarus. As Gazprom sees it, both countries have at times blackmailed Russia, either cutting off or threatening to cut off the flow of oil or gas to Western Europe. That is the primary reason that Gazprom and Putin have worked with Germany and particularly Chancellor Gerhard Schroeder to support the construction of a gas pipeline under the Baltic Sea directly from Russia to Germany. Now called the "Nord Stream" pipeline, Gazprom has a 51 percent share in the consortium that is building and operating the pipeline. Two German companies, E.ON and Wintershall, a wholly owned subsidy of BASF, each initially had 24.5 percent of the remainder.[52]

This pipeline has been surrounded by controversy, both within Germany and in Eastern Europe. For the Germans, it was embarrassing to discover that Gerhard Schroeder, after having been so outspoken in support of building this bypass Nord Stream pipeline while chancellor, became the chairman of its board immediately upon being voted out of office. (At the time it was called the North European Gas Pipeline Company [NEGP].)[53] The embarrassing part was that for this relatively cushy, figurehead job, Schroeder was to be paid an annual salary of $300,000.[54] Moreover, not only had he been the main sponsor of such a pipeline within Germany but days before he left office, the German government offered to act as a guarantor for a 1 billion euro loan which a consortium of German banks was prepared to offer to finance the project.[55] This would make such a loan more attractive to the banks and thus result in a lower interest rate.

Critics questioned why Gazprom, one of the world's wealthiest corporations, needed such a guarantee. In the words of Guido Westerwelle, a German opposition party leader, "This affair stinks terribly."[56] Many Germans saw Schroeder's appointment as chairman of the pipeline consortium not only as a blatant conflict of interest but also as outright prostitution. They called for Schroeder's resignation from the chairmanship. Schroeder denied that he even knew such a loan guarantee had been offered and Gazprom insisted that it did not need a loan, much less a loan guarantee.[57] Schroeder refused to resign but the incident did little to improve either his or Gazprom's image, much less its transparency.

Schroeder's role in this was not a black and white matter. It certainly made sense to argue—as he did—that Germany should diversify its sources of energy supply so that it would be less dependent on the Middle East. Moreover, with the continuing unrest in the Middle East, supplies coming overland by pipeline from a continental neighbor seemed a safer bet than supplies shipped from the Persian Gulf and through the Suez Canal.

From the environmentalists' point of view, there was also something to be said on behalf of such a pipeline, at least in part. Once he was assured of supplies of natural gas from Russia, Schroeder ordered all of Germany's nuclear energy plants to be closed down by 2021.[58] In 2004 nuclear reactors generated 30 percent of Germany's electricity, so to do away with nuclear energy would mean that Germany would have to generate significant quantities of electricity with other fuels.

The downside of these otherwise praiseworthy initiatives was that this pipeline would increase Germany's dependency on Russia. Would Russia adhere to its contracts, even if there should be a future political disagreement? Because the Soviet Union had held to its contract commitments with Germany and others in Western Europe even during tense periods of the Cold War era, those favoring such dependence on Russia felt the risk was worth taking.[59]

From the American point of view, however, concerns were raised by the knowledge that both the Soviet Union and Russia had a history of ignoring contractual agreements with a number of non-West European countries. Of course, halting the flow of gas to Germany would be a much more momentous matter than cutting off the flow to Ukraine, but if there were a serious enough dispute, the Russians might do just that. While some U.S. as well as European policy makers began to worry that the showdown with Ukraine in January 2006 was a forerunner of many more such incidents, the German public was much more titillated by the scandal that centered on their prime minister. It stemmed from Schroeder's appointment as head of the Nord Stream Pipeline and Schroeder's brazen efforts subsequently to obtain a gag order from a Hamburg court halting criticism of his backing of the project. Nor were skeptics reassured in August 2007 when Gazprom acknowledged that the cost of building the pipeline would be 50 percent higher than initial promises. Moreover, there was also a strong likelihood that there would be yet other costs in the future.[60]

Adding to the notion that the pipeline project had become a honeypot for payoffs and a form of apparatchik nepotism, Matthias Warnig was appointed managing director of Nord Stream, the company that would build and operate the pipeline.[61] Warnig, who at the time was board chairman of the Russian branch of Dresdner Bank, had worked with Putin in East Germany during the 1980s when both were intelligence agents: Warnig a captain in the Stasi, the East German Secret Police (or the East German Ministry of Foreign Trade, as his official biography describes it), and Putin a lieutenant colonel in the KGB.[62] Their paths crossed again in St. Petersburg when Putin was put in charge of the mayor's office for foreign economic relations. Warnig negotiated with Putin for an operating license in St. Petersburg for

Dresdner Bank, and it became the first foreign bank to receive such permission to operate in St. Petersburg.

Having friends in the U.S. White House is not that much different, but clearly, friendship with Putin does pay. Schroeder and Warnig are the best examples. Before he became managing director of Nord Stream, Warnig was also nominated to Gazprom's board of directors. It did not hurt the Putin-Warnig relationship that in 1993 Warnig stepped in to pay for an emergency private plane flight to a German hospital for Putin's wife after she had a serious auto accident.[63] Warnig also helped finance the living expenses for Putin's two daughters while they were studying in Germany.

While the Germans favored the Nord Stream project, the East Europeans—particularly Ukraine, Belarus, and Poland—opposed it. They were not only worried that Nord Stream would eliminate their chokehold on such shipments, but they were also concerned that they would lose substantial transit fees. Besides, there were legitimate fears that the construction of such a pipeline would cause further damage to the ecology of the Baltic Sea, which was already seriously polluted. Incidentally, other neighboring states, including Latvia, Estonia, and Lithuania, worried that any underwater construction might trigger the explosion of the numerous mustard gas containers the Germans had dumped into the Baltic at the end of World War II.[64] Such concerns ultimately have forced a review by Nord Stream of the pipeline's environmental impact. Sweden insists that Nord Stream must have the approval of all the countries whose territory will be traversed by the pipeline. For that reason, Estonia has extended its claim to sovereignty over the territorial waters from three to twelve nautical miles, which means that the Russians will now also need Estonia's permission. However, Estonia has rejected Nord Stream's request to conduct a survey of the Baltic seabed in Estonia's offshore economic zone. With such complications, construction has been postponed for at least one year.[65]

The Poles had their own fears. Given their history, some saw the pipeline agreement as a conspiracy by Germany and Russia to gang up on their immediate neighbor. For a time it seemed to be a replay of the Rapallo Pact of 1922, which some argue helped Germany re-arm after World War I. However, after Donald Tusk became prime minister in late 2007, he and German chancellor Angela Merkel worked to reassure Poland, and Ms. Merkel also offered to divert gas to Poland from the German pipeline should there be a need to do so.[66]

As if all this were not enough to cloud the project, knowledgeable specialists from Sweden have told me that the Swedes are also opposed to such a Baltic Sea pipeline. Their opposition goes beyond the environmental concerns of their Baltic neighbors. The Swedes are worried that the Russians will use such a pipeline to install underwater listening and eavesdropping equipment.[67] This would allow the Russians to monitor the commercial traffic as well as Swedish military communications, just as the Swedes presently use similar equipment to monitor communication within Russia. Officially Sweden is neutral and not a member of the NATO pact, but whether a member or not, Sweden nonetheless maintains close intelligence connections with NATO headquarters. Officially, Sweden argues that its permission is needed for such a pipeline because it will be built within what Sweden considers its exclusive economic zone.[68] Its ostensible concern is that because the Baltic is such a shallow sea, the pipeline will serve as a barrier to existing poor water circulation and thereby increase pollution within the already vulnerable seabed. Should the Russians go ahead with the pipeline's construction, some Swedes have told me that the Swedish military have drawn up plans and are fully prepared to sabotage the pipeline if and when it is built.[69]

OPPOSING CASPIAN OIL AND GAS ALTERNATIVES

While Russia has had to fight off efforts by countries bordering the Baltic Sea who want to prevent the building of a direct Russian-German gas pipeline, there has been a somewhat similar struggle over a Caspian Sea pipeline, only this time Russia seeks to outmaneuver and discourage efforts by Western companies, Central Asian producers, the European Union, and the United States. As we saw, they seek to build a non-Russian alternate undersea route for both gas and petroleum pipelines from producers operating within the Caspian Sea basin. This gas pipeline would be built under the Caspian Sea, and depending on the particular proposal, link up with either or both Turkmenistan and Kazakhstan. The pipeline would then come ashore in Baku and flow through to Ceyhan, Turkey, on the shore of the Mediterranean Sea. The Russians have warned that they would oppose such an underground gas pipeline until "the legal status of the Caspian Sea" is resolved. After the breakup of the USSR, three new countries— Azerbaijan, Kazakhstan, and Turkmenistan—all began to claim underwater rights, some of which had previously been held by either Iran or the USSR (until 1991 the only two countries with Caspian Sea shoreline).[70] Not surprisingly, Russia evidenced no such concerns as it sought to build its own gas pipeline under the Baltic Sea.

Because the petroleum pipeline did not involve construction under the Caspian Sea, there was no petroleum pipeline proposal for the Russians to oppose. The Trans-Caucasian petroleum pipeline was all overland and on non-Russian land. As we saw, prompted by the United States, BP and some other producers in the region constructed a petroleum pipeline from Baku in Azerbaijan through Tbilisi, Georgia, on to Ceyhan, Turkey, on the Mediterranean Sea. This petroleum pipeline satisfied three needs. First, it provided an outlet to the Mediterranean and on to Europe for non-Russian petroleum producers so their petroleum shipments did not have to pass through the narrow and therefore dangerous nineteen-mile long Bosporus Strait in Istanbul. Second, it also offered an alternative to the Caspian Pipeline Consortium (CPC) that transports petroleum from Kazakhstan—a country with huge production potential—to Novorossiysk in Russia on the Black Sea. Finally, it also provided a right of way for the parallel natural gas pipeline that opened in December 2006 and is intended to link up with the NABUCCO gas pipeline further to the west, which, as stated, was also designed to provide an alternative to the Gazprom pipeline monopoly.

To ensure that the Baku-Tbilisi-Ceyhan oil pipeline will not be profitable, Putin has done his best to provide cheaper alternatives through overland Russian routes. For that reason, in March 2007 he agreed to promote the construction of the Trans-Balkan Oil Pipeline from Burgas, Bulgaria, on the Black Sea, to Alexandroupolis, Greece, on the Aegean Sea, which would also bypass the Bosporous. To guarantee that this new Trans-Balkan Oil Pipeline will attract enough volume, Russia consented to the expansion of the semi-privately owned Caspian Pipeline Consortium (CPC) from Kazakhstan, which the oil companies need badly; however, there was a condition: in exchange for allowing the expansion of the CPC pipeline, the companies must also agree to use this proposed Trans-Balkan Oil Pipeline—which of course will reduce the supplies available to ship through the BP-backed pipeline alternative through Azerbaijan and Georgia. In what seems to be an effort to intimidate them into supporting these Russian projects, the CPC members have been charged with failing to pay $290 million in back taxes to Russia.[71] When built, this Burgas-Alexandroupolis pipeline will be the first pipeline within the European Union itself that will be controlled by a Russian state agency. Expanding into Europe, Russia will hold a 51 percent share and Bulgaria and Greece each 24.5

percent. These moves are designed to counter the EU's efforts to reduce its dependency on Russian oil.[72] Even though construction will not be completed before 2011 at the earliest, use of the Burgas-Alexandroupolis route will help frustrate Western efforts to send petroleum from the Caspian region on through Georgia across the Black Sea and then to the Odessa-Brody pipeline in Western Ukraine. The Odessa-Brody pipeline, built in 2001, was designed to take Caspian oil from Odessa north to Poland and the EU. Because the volume and shipments were too low to make this profitable, the Russians instead arranged to reverse the flow and send petroleum south from Brody to Odessa and on to Burgas, Bulgaria. While efforts continue to find enough petroleum to make it possible to send non-Russian petroleum north, for a time at least the Russians seem to have prevented Western companies from bypassing a Russian chokehold.[73]

PIPELINES IN ASIA

Russia's pipeline diplomacy is not limited to Europe. Asia's dynamic economies are also important markets for Russian energy exports. Russia has enormous potential for oil and gas development off the island of Sakhalin and for gas at Kovykta in East Siberia. Japan, South Korea, China, possibly India, and even the United States are all potential customers. On a 2006 visit to China, Putin indicated that Russia would build two gas pipelines to China, one from East Siberia and one from West Siberia. As a measure of how important Asia is expected to become to Russia, at a September 2006 meeting, Putin told a group of us that Russia's energy exports to Asia would increase from 3 percent of the country's total energy exports in 2006 to 30 percent by 2012. He did not indicate where, if at all, its energy exports would be cut back, but the implication is that at least the share of energy exports—if not the actual volume—destined for Europe would be smaller.

The sale of so much petroleum and natural gas is predicated on the assumption that the parties can agree on prices (China is a particularly tough negotiator) and that an agreement can be reached as to who will build and operate the pipeline. The latter should be a simple matter, but nothing is simple in these dealings. In the case of Kovykta, for example, Gazprom insists that TNK-BP, which has developed the field, cannot build its own pipeline. Rather, it can only transport its gas through the Gazprom pipeline, and Gazprom will do that only when Gazprom is allowed to have a major equity share of the project. As we saw, Gazprom also muscled out Royal Dutch Shell from dominant control in the Sakhalin II project in much the same way. (We will see that when the sides were reversed, the Russians were quite unhappy when the EU proposed that as a producer of natural gas, Gazprom should be precluded from owning and controlling the gas distribution pipelines as well.)

Given the distances involved, constructing a pipeline to China is a major engineering challenge. But Kazmunai, the Kazakh state oil company, has already completed a 970-kilometer petroleum pipeline from Central Kazakhstan to Xinjiang in northwest China, which it opened in May 2006.[74] Both countries have agreed to extend the pipeline to western Kazakhstan near the Caspian Sea oil fields. The CITIC group of China, along with the China National Petroleum Corporation, has spent over $6 billion to buy up shares in Kazakh oil fields.[75] The Russians are engaged in similar pipeline construction efforts, including one project that would involve transporting Russian oil via the Kazakh pipeline on to China.

Shipping petroleum or gas to China is also an important part of Gazprom strategy. Whenever Europeans try to reassure themselves that they need not fear that the Russians will use energy to bully them because the Russians need Europe to buy its gas as much as Europe needs to buy the gas, Putin runs off to Asia with promises that even though it will be very costly, Russia will ship gas from fields the Europeans assumed had been set aside for their use.[76] And if China refuses to pay Russia's prices, Putin knows there are customers in Japan and South Korea who will. Moreover, once it does start to accept Russian gas, China is as likely to become as dependent on it over time as the Europeans and to find itself becoming as vulnerable as Europe to the possibility of political pressure and on occasion blackmail.[77]

Of course, the Russians insist they will never, ever, allow political disagreements to interfere with contractual agreements. According to Alexander Medvedev, deputy CEO of Gazprom, "For us contracts are like a Holy Bible."[78] He has been echoed by Igor Shuvalov, the economic adviser to President Putin, who told the *Financial Times* that Russia "did not like" the fact that the European Union felt it necessary to diversify its energy suppliers. "We've always said the same thing; we are the most reliable supplier, in any circumstances, for the European market. The most reliable. Like it or not, even if people question it. Europe will never have a more reliable supplier of energy than Russia."[79] In much the same spirit, Sergei Karaganov, chairman of the Russian Council on Foreign and Defense Policy, criticized U.S. Senator

Richard Lugar for referring "to Russia as an unstable country in talking about NATO's energy security. . . . Indeed, Russia has now a reputation over several years as a reliable partner for the West in terms of supplying energy resources."[80] Putin himself has also stressed Russia's reliability. At a press conference on February 1, 2007, Putin charged, "We are constantly being fed the argument that Russia is using its current and emerging economic levers to achieve its foreign policy goals." He insisted, "This is not the case. The Russian Federation has always abided by all of its obligations fully and completely, and it will continue to do so."[81]

Were it only so. Admittedly, the Soviets held to their supply contracts with NATO countries, like Germany and Italy, through the worst of the Cold War. But ironically, other countries, some of which were one-time Soviet allies or part of the USSR itself, were not so lucky. Although Putin and his associates may not find it in either their Soviet or Russian history books, as we saw earlier in Chapter 2, there have been almost a dozen instances when both petroleum and gas deliveries were suspended for political or economic reasons in mid-contract by both Soviet and Russian energy exporters. Behavior like this, and the denials that such things ever happened, should make those dependent on Russian gas deliveries very nervous.[82]

OGEC

For a time, in 2006–2007, there was debate as to whether Russia would be able to create a gas counterpart to OPEC. Putin visited all the usual suspects. He discussed such an arrangement with leaders of Iran, Algeria, and Qatar, the most likely participants in such an organization. Even earlier in 2002, Putin had proposed that Russia and the Central Asian gas producers explore the possibility of creating an "alliance" to coordinate the transportation of their natural gas, a trial balloon he soon dropped.[83]

But the gas market is different from the petroleum market so that an OPEC-like organization, an OGEC (Organization for Gas Exporting Countries), would not make sense. Unlike petroleum producers, gas producers cannot easily shift their deliveries around to other countries. To the extent it is effective, OPEC must be able to induce restraint among producers of petroleum from doing just that. This generally means reducing the supply of petroleum on the market so that at existing prices there will be more demand for petroleum than producers are willing or able to supply. But for such a tactic to be effective, each OPEC member must limit its daily production so as to hold down competitive pressures and price cutting. This tightens the market and frequently leads to an increase in crude oil prices. By contrast, since most natural gas is delivered to its customers by pipeline, there is usually no other viable or affordable source of supply available. Some say that LNG could serve that purpose, but producing and delivering it is very expensive—so much so that producing and selling LNG is viable only when the parties are also willing to sign long-term contracts. This explains why there is as yet only a limited international spot market for LNG, which contrasts with the petroleum spot market where many last-minute purchases can easily be arranged.

Because of spot market pricing in the buying and selling of petroleum and the absence of anything similar to the natural gas market, oil prices, unlike gas prices, tend to be uniform around the world. According to a study by Richard J. Anderson at the George C. Marshall Center in Garmisch, Germany, because there is no such flexibility or ability to substitute suppliers in the natural gas market, prices for natural gas will vary as much as 31 percent from place to place on any given day. In the vocabulary of economists, there is very little room for arbitrage in world gas markets.

While Putin's discussions with Algeria, Iran, and Qatar are unlikely to result in the actual formation of an OPEC-like organization, the Gas Exporting Countries' Forum (GECF), which was formed in 2001 and has met only sporadically, may attempt to increase the sharing of information on prices and technology, but not much more. Russia has refused to join OPEC because it did not want to feel constrained by the decisions of such a coordinating group in the way it sells its petroleum. Unless it can work out an arrangement assuring that it will always be able to dictate GECF policy, it seems unlikely that Russia would be willing to accept decisions about how and when it can sell its natural gas.[84] In actual fact, given the difference in the way gas is delivered, a gas supplier is less likely to need an OPEC to exercise economic and political leverage. Unlike the petroleum markets, which need coordinated behavior among a substantial number of producers to control price and supply, a supplier of natural gas is more likely to have a monopoly relationship with its customers. This is the kind of market OPEC tries to create, but to be effective, it must mobilize a concerted effort by more than a dozen producers. By contrast, because it already is the sole supplier of gas to many of its customers, Russia is effectively a one-country OGEC: an Organization of a Gas Exporting *Country*, in the singular.

Of course, Russia is not the only source of Europe's natural gas. Norway and Algeria are major suppliers, and the United Kingdom and the

Netherlands can supplement output. But reserves in all these countries are being depleted. While they are all connected to the pipeline network, by 2006 there was very little excess capacity available if Russia, as the major supplier, were suddenly to suspend its deliveries to its customers.

Table 6.2 shows just how important Russian gas is to Europe. Over a quarter of all the gas consumed there comes by pipeline from Russia. In the extreme case, Finland and the Baltic countries depend on Russia for 100 percent of their gas. But Germany, which buys a larger volume of natural gas from Russia than from any other country, depends on Russia for more than 42 percent of its imports. Russia provides 38 percent of its overall gas consumption. This is despite initial promises to limit dependence on Russian gas to 30 percent of overall consumption. As the reserves of the other suppliers are drawn down, dependence on Russia is expected to increase. If there should be any break in the flow, neither Norway nor Algeria can do much to make up the difference. Although they are at the other end of the pipeline, even Italy and France each depend on Russia for more than 30 percent of their imports. This

TABLE 6.2   Europe's Reliance on Russian Gas (Bill. Cubic Meters), 2004

	Total consumption	Total imports	Imports from Russia	% of total consumption	% of imports
Europe	526	372		26	
Germany	97	91	36	38	40
Italy	81	68	21.6	26	32
Turkey	23	22	14.5	64	66
France	45	45	13.3	25	30
Poland	14	10	6.3	42.5	63
Austria	9	8	6	65.7	75
Hungary	14	11	9	66	82
Czech Republic	9.6	9.5	6.8	74.6	72
Slovakia	6.6	6.4	5.8	97	91
Finland	4.6	4.6	4.6	100	100
Estonia	0.97	0.97	0.97	100	100
Latvia	1.75	1.75	1.75	100	100
Lithuania	2.93	2.93	2.93	100	100

Data from natural gas information, International Energy Agency (OECO), 2005

heavy dependency partly explains why Gazprom was able to convince the French and their gas company, Gaz de France, to allow Gazprom to take over the internal pipeline delivery of three billion cubic meters of gas directly to individual French households. (See Table 6.1.) This of course gives Gazprom even more power. The Italian company Eni, a long-time trading partner of Gazprom and Enel, has also agreed that by 2010, Gazprom can sell up to 3 billion cubic meters directly to Italian households and factories.[85] In exchange, Eni was allowed to buy gas-producing assets within Russia. The Italians have sought to acquire Gazprom's 19 percent share of ownership in Novatek, which is now Russia's second largest producer of gas.

Since Gazprom exports so much gas to Germany, it has made a special effort to integrate itself into Germany's domestic distribution system. It has become closely connected to Germany's three major gas supply companies, E.ON, Wingas, and Wintershall. The latter is a wholly owned subsidiary of BASF, a major multinational chemical corporation, which is involved in a maze of interlocking directorates. For example, Wintershall and Gazprom are partners in Wingas (Win and Gas), in which Gazprom has a 50 percent equity less one share. Beyond that, Gazprom, Wintershall, and E.ON have created another joint venture to develop the Yuzhno-Russkoye gas field in Russia. In this joint venture, E.ON and Wintershall each have one share less than a 25 percent equity. As for Wingas, it originally had a 49 percent interest in Nord Stream, the proposed Baltic gas pipeline. In this case Gazprom has the majority 51 percent portion. Wingas will tie in Nord Stream to the internal German gas grid. E.ON, which was created in June 2000 as a joint venture by the German companies VEBA and VIAG, was allowed to acquire 24 percent of the Nord Stream project from Wingas and BASF.

While Gazprom continues to hold 51 percent of Nord Stream, the German companies have had to spin off some of their shares to Gasunie, a Dutch company. In November 2007, Putin and Dutch premier Peter Balkenende announced that Gasunie had acquired a 9 percent share in the pipeline. This forced both E.ON and BASF/Winstershall to reduce their equity from 24.5 percent each to 20 percent each. Gazprom kept its 51 percent share. Equally important for Gazprom, part of this partnership arrangement includes an option for Gazprom to purchase 9 percent of the Balggand-Bacton pipeline that connects the Netherlands to Great Britain, an access Gazprom has long sought.[86]

If this were not confusing enough, Gazprom in turn can purchase up to 25 percent in E.ON. The German firm Ruhrgas, which in 2003 was bought up by E.ON, owns 6.5 percent of Gazprom. So here is how things stand. E.ON owns Ruhrgas which in turn owns part of Gazprom, and Gazprom can buy up part of E.ON. This is like a dog trying to grab hold of its tail. To top it off, Bergmann Bruckhard, chairman of the Management Committee of Ruhrgas, is one of the few foreigners who is on the Gazprom Board of Directors.[87]

All of this is very reminiscent of the way Soviet authorities designed their overseas trading and banking networks during the Soviet era. Each Soviet overseas corporation owned shares in almost all their fellow overseas corporations. This was done to mask responsibility while creating the appearance that the Soviet corporation had private shareholders and owners like other corporations.[88] In sum, although who owns whom is convoluted and as hard to follow as the pea in a sidewalk shell game, the Germans are very much involved with Gazprom, and Gazprom, in turn, has become an important player in Germany.

Direct access to the French, German, and Italian consumers allows Gazprom to earn a higher margin on its sales. It also gives it greater control over the source of the gas sold within these countries and again is a way of excluding other suppliers. The effort to gain dominant control is part of Gazprom's long-term strategy. In addition to France, Italy, and Germany, Gazprom has either already succeeded or is trying to gain control of internal gas pipelines and distribution systems in Belarus, Ukraine, Georgia, Moldova, Switzerland, Austria, Finland, Turkey, Hungary, Greece, Latvia, and Lithuania, where Gazprom now owns 34 percent of Lithuania's pipeline grid company, Lietuvos Dujos.[89]

THE RUSSIANS ARE COMING

Alert to the strategic control Gazprom would gain from internal pipelines and distribution systems, some gas distributors have become wary of allowing Gazprom to make such inroads. After Gazprom began to explore the possibility of buying up Centrica, Great Britain's largest gas distributor, the *Financial Times* published an editorial entitled, "Your Local Gazprom," warning British consumers that they might find themselves subject to Kremlin control. It could have added that Gazprom has also attempted to gain control of some British electricity-generating facilities as part of a swap arrangement with the German company, Ruhrgas.[90] The paper acknowledged that foreign companies from the United States, Germany, and France were also taking control of energy assets in the United Kingdom, but given Russia's past record it was concerned that the possibility of Russian control brought with it other negative "geopolitical factors to which unfortunately Gazprom is inherently prone."

Such concerns go beyond Great Britain and the European continent. Alexander Medvedev, deputy CEO of Gazprom, for example, has implied that some day Gazprom might create a joint venture that would distribute gas in China's domestic market.[91] For that matter, there is nothing to prevent Gazprom from making a similar investment in U.S. gas companies. LUKoil's purchase of Getty Oil's filling station network is a precedent. In addition, several Russian metallurgical companies have already acquired a variety of U.S. steel and nonferrous metal companies, including the only U.S. producer of platinum and palladium.

Not surprisingly, the Russians do not take kindly to suggestions that Europeans should be wary of allowing Russian companies to expand beyond their borders. After Alan Johnson, British minister of Trade and Industry, insisted that England would block Gazprom from taking over Centrica, the parent company of British Gas, Alexei Miller, Gazprom's CEO, warned that "attempts to limit Gazprom's activities in the European market and politicize questions of gas supply, which in fact are of an entirely economic nature, will not lead to good results."[92] His response not only conveys Russia's sensitivity over efforts to exclude it; it also reveals his insensitivity. Miller angrily chastised the British and Europeans for acting for political reasons while he apparently failed to realize that to Western observers, it is the Russians, even more than Western governments, who place political considerations ahead of commercial and economic considerations.

7

Russia

The Unrestrained Super Energy Power

THE ENERGY GIANT REAWAKENS

In 1999, as petroleum prices began their climb from $10 a barrel to over $100, memories of the 1998 financial meltdown and its impact on Russia quickly faded. Fortunately for him, Vladimir Putin's selection as prime minister in August 1999 and four months later his appointment as acting president coincided with the recovery in petroleum prices. The increase in oil prices would probably have triggered an economic recovery even if Boris Yeltsin had still been in power. Nevertheless, Putin did what he could to take advantage of that recovery in oil prices.

Putin's first priority was to prevent any further deterioration in Russia's political and economic situation. In the aftermath of the August 17, 1998, economic crisis, it was difficult to see anything but a continuing deterioration in the economy. The banks remained closed, and because of the sharp drop in the value of the ruble more and more businesses, especially those run by foreign companies or dependent on imported components, closed their doors as well. Many of the country's most talented people lost almost all their savings. No wonder so many simply emigrated to the West.

The devalued ruble, however, along with the gradually rising price of oil, proved to be that proverbial blessing in disguise. A cheaper ruble

FIGURE 5 How Price, Not Putin, Affected Oil Production

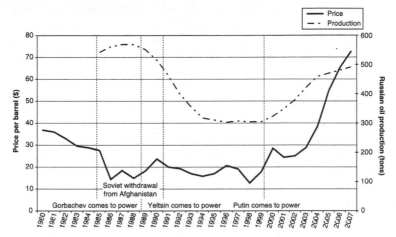

Crude oil "price per barrel" is U.S. dollars per barrel, Brent; Russian production is in tons per year. Sources: Price data from BP Statistical Review of World Energy June 2007, p. 16; 2007 price from Energy Information Administration, Weekly Petroleum Status Report (week of February 8).
Production data from 2006 BP Statistical Review of World Energy; 2007 production value from Goskomstat (Russian official statistics).

made imports more expensive, so Russians began to buy goods made in Russia instead. The drop in imports may have hurt Russian consumers and businesses that depended on imports, but it was a windfall for Russian manufacturers who suddenly had the domestic market to themselves. This windfall explains why for the whole of 1999 industrial production increased 6.4 percent over 1998 and by 10 percent in 2000 over 1999. Even though Putin was appointed prime minister in August 1999, five months after industrial production started to increase, he took office the same year the economy began its recovery. So it is easy to see why, for many, including *Time* magazine, which named him "Man of the Year" in 2007, Putin was the reason for the turnaround.

The causes and effects of Putin's actions in the political sphere are harder to pinpoint. Putin began to move against what he considered the excessive number of political parties in the country. With more than one hundred existing parties, too many, he insisted, were no more than vehicles for individual ego building and petty feuds. He felt much the same way about the media. As he saw it, several of the oligarchs were using ownership of their TV networks primarily to attack each other rather than advance the interests of the state. Putin never bothered to mention that his targets were almost exclusively those TV networks that targeted him for criticism. It hardly advances the interests of the state, much less the cause of democracy, when a country's leader like Putin can single-handedly determine which TV networks should be allowed to operate and which should be closed down; nonetheless, he did have a point.

So whether Putin can be considered a supporter of democracy is not a simple black or white matter. Certainly he has no doubts. After Gerhard Schroeder (not one who could be described as unduly critical of things Russian, especially Putin) described Putin as "a pure democrat," a reporter asked Putin if the characterization was accurate; Putin responded, "Of course I am, absolutely. . . . The problem is I'm all alone, the only one of my kind in the whole world." To prove his point, he criticized the United States and Europe for Guantanamo, detention without trial, the home-less, and rubber bullets and tear gas against European demonstrators. In his view, no one else seems to care. As he put it, "There is no one else to talk to since Mahatma Gandhi died," all said with a straight face.

Of course, not everyone would agree, even within Russia. Grigory Yavlinsky, head of Yabloko, one of the more Western-oriented political parties, put it this way in an interview in the July 15, 2006, issue of the *Economist*. "Boris Yeltsin took mistaken steps in the right direction toward democracy; Putin took correct steps in the wrong direction toward an authoritarian petro state."

Perhaps the most controversial step he took while he was still prime min-ister in 1999 was to disavow the informal cease-fire accepted by Yeltsin in 1996 and order Russian troops back into Chechnia. In effect, he launched the second Chechen war of the twentieth century. Putin did this in retalia-tion for the bombing of a series of apartment houses in Moscow and elsewhere. He ordered his troops to invade, even though there was no evi-dence that the Chechens had actually planted the bombs. He also felt it nec-essary to respond to the invasion of neighboring Dagestan by an extremist Chechen group. Critics of Putin such as Boris Berezovsky insist that the bombings were actually a provocation set off, in fact, by the FSB (formerly called the KGB) itself. Whatever the provocation, Putin used the war to rally the country to fight what he saw as a potentially disastrous terrorist threat.

THE NEW ECONOMIC IMPERIALISM

Putin's most significant contribution to Russia's economic and political renaissance, however, was his adoption of the notion of national champions. It was his way of merging state interests with private sector capabilities. Putin correctly understood that Russia had little in its economic and business arsenal other than its energy and mineral resources. Skillfully used, Russian petroleum, gas, and other exotic minerals could be manipulated to advance state interests. But under Yeltsin, the preponderance of the country's raw material reserves had been turned over to individual oligarchs and their corporations who used the country's energy and metal resources to advance their own interests and profits. From Putin's point of view, this was outrageous. In a June 1, 2007, press conference, he expressed "regret" that in the early 1990s Russian officials had allowed such transfers to take place, actions for which they should have been put in prison.[1]

Even so, Putin insisted that he supported privatization. In a June 4, 2007, press conference, he boasted, "We have completely privatized our oil sector and we now have only two companies with state participation. Gazprom already has 49 percent of its shares on the market and according to our calculations, more than 20 percent are now in foreign hands. . . . The other company Rosneft carried out an IPO [initial public offering of stock] and as you know has sold part of its shares." What mattered, however, was not who actually owned the shares but whether the managers of these companies acted as agents of the state and adhered strictly to the goals set out by Putin and other senior state officials as if they were wholly owned by the state.

Putin's embrace of the concept of privatization notwithstanding, he nonetheless set out to reassert the state's interests by either renationalizing the country's corporations or by applying subtle—and sometimes not so subtle—intimidation to convince those corporations that they should temper profit considerations in favor of advancing what Putin had decided were the country's geopolitical or strategic goals. While Putin may have been one of the most recent world leaders to openly espouse such a notion, he is not the only one to do so. The president of France, Nicolas Sarkozy, began to call for the same type of initiative when he was minister of the Interior, an idea earlier supported by Charles de Gaulle when he was France's president and even earlier by Jean-Baptiste Colbert, who was King Louis XIV's financial controller in the seventeenth century.

As market conditions tightened in the early twenty-first century, the idea of Russian national champions became particularly attractive. The advent of China and then India as voracious consumers of energy and metals put Russia, with its abundant resource deposits, in a strong bargaining position. Taking advantage of these changes in market conditions, Putin skillfully utilized Russia's gas and oil potential to advance its economic and political agenda. At times his efforts seemed little different from what the Soviets used to call "economic imperialism." The difference in this case, if there is any difference, is that in the pre–World War I era, most of the "capitalist" corporations controlling such resources were privately owned. But private or state owned, after they established a foothold in a foreign country, they pressured their home country to help them maintain their interests. In Putin's Russia, most national champions are either wholly or predominantly state owned, although some, such as Surgutneftegaz, have no or limited state ownership. Whether public or private, these national champions, actively encouraged by the state, seek to dominate foreign markets, just as companies did in the pre–World War I era. Usually the corporation takes the first step to establish a foreign presence, but on occasion the state has acted first and only afterward was the national champion brought in to carry out a state-to-state agreement (see Tables 6.1 and 7.1).

This economic imperialism—as Lenin would have labeled it—is not necessarily limited to the outside world. A reverse form of economic imperialism has essentially taken place within Russia itself. As we have seen, after both the 1917 Revolution and the 1991 breakup of the USSR, the government found it necessary to offer concessions to foreign energy companies because Russian companies were unable to exploit the country's oil deposits on their own. Unable to master the drilling challenges in extreme circumstances, particularly offshore, the Soviets in 1917 and the Russians in 1992 found it necessary to bring in foreign technicians. In the 1990s, the government even agreed to accept production sharing agreements (PSAs) with significant tax concessions for foreign companies—the sort of policy followed by much poorer and smaller third world nations. But once Russia and its industries recovered enough economically to do without such help, the state authorities either disregarded contractual agreements or found environmental loopholes or instances of tax evasion that they used to claim contractual violations, as they did with Shell at Sakhalin and BP at Kovykta.[2]

The bankruptcy and renationalization of Yukos, as we have seen, was an extreme example of how the state will resort to extreme measures to regain control of a private enterprise. To the victims, of course, it felt like a form of domestic economic imperialism. But this renationalization sent a clear message. Rare is the corporate chief executive officer, Russian or foreign, in Russia who today dares to defy state edicts or wishes; they all realize that if the state prosecutor wants to, he can find something they have done that was illegal. As explained by Boris Berezovsky, the exiled former oligarch behind Sibneft, Aeroflot, and several other previously

TABLE 7.1 Russian Petroleum Company Expansion Abroad

	Company	Filling Stations	Oil Exploration and Production	Fuel Distribution	Oil Refinery	Russian Company
Belgium	Conoco petrol stations	×				LUKoil
Bulgaria	Neftochim	×			×	LUKoil
Czech Republic	Conoco petrol stations	×				LUKoil
Finland	Teboil / Suomen Petrooli	×		×		LUKoil
Greece				×		LUKoil
Hungary		×				LUKoil
Kazakhstan	Nelson resources		×			LUKoil
Lithuania	Mazelkiu nafta				×	Yukos
Poland	Conoco petrol stations	×				LUKoil
Romania	Petrotel Lukoil	×			×	LUKoil
Serbia/Montenegro	Beopetrol	×		×		LUKoil
Slovakia		×				LUKoil
Slovenia		×				LUKoil
Ukraine		×			×	LUKoil
USA	Getty petroleum marketing	×		×		LUKoil

Sources:
Financial Times, January 24, 2007, p. 3.
Russian Profile, March 2006, p. 30.

privatized entities, everyone in business in Russia must necessarily have violated the law at one time or another. Given the helter-skelter, often contradictory, nature of the privatization process and the absence of any well-established interpretation of Russian legal codes, it was impossible to adhere to the letter of the law and operate profitably. A survey among businessmen and women conducted in 2001 found that only 15 percent of those interviewed claimed they could operate legally. The remaining 85 percent said that of necessity they have had to cut corners.[3] As a result, almost everyone operates with the knowledge that if they step out of line or cross the wrong person, they can face crippling charges from the prosecutor general's office. Once the prosecutor general makes such charges, he can then freeze a company's bank account, after which it cannot pay its bills and almost inevitably cannot survive.

POLITICIANS FOR SALE

Just as the state and its national champions can force out whomever they choose, they can also use their resources to buy up priority projects or personnel. The most glaring instance of this was the way Gerhard Schroeder prostituted himself for the Nord Stream gas pipeline designed to link Russia and Germany under the Baltic Sea.

As embarrassing as Schroeder's complicity was to most Germans, he is not the only one to be wooed or seduced by Russia's energy money. About the same time, Donald Evans, a close friend of President George W. Bush, was offered a somewhat similar opportunity. Evans, like Bush, also has a background in the Texas oil business. (Reportedly, he was more successful at it than the future president.) Later, Evans served as chairman of the Bush-Cheney 2000 presidential campaign, after which Bush brought him to Washington in January 2001, as his first Secretary of Commerce. Attempting to take advantage of that close relationship with President Bush, in December 2005, a year after Rosneft seized Yuganskneftegaz from Yukos, President Putin offered Evans what Putin referred to as "a top job" at Rosneft.

In a May 2006 meeting with President Bush, the president told a small group of us that he had heard about the offer to Evans directly from Putin, who told Bush that he was doing this "as a favor" to President Bush. The president said he was puzzled as to why Putin thought this should be regarded as "a favor." From the outside it was anything but. Accepting this offer from Rosneft would have served to legitimize Rosneft's takeover of Yuganskneftegaz, which had been

Yukos's most valuable asset. If he had any doubt as to how compromising his acceptance would have been, a December 19, 2005, editorial in the *Wall Street Journal* warned Evans that the public would perceive his taking the job as a sellout and urged him to reject the offer. After thinking about it for two weeks (suggesting that he must have been tempted), Evans eventually turned it down, thereby sparing himself the embarrassment that followed Schroeder's appointment.

Not everyone in the United States was so principled or so forewarned. Shortly after his daughter Karen was hired for $500,000 by ITERA, the Turkmenistan-Ukraine trading company headquartered in Jacksonville, Florida, Curt Weldon, a Republican congressman from Pennsylvania, became ITERA's public advocate. Those connections in part explain how ITERA, a rich Russian energy company, managed to apply for and receive an $868,000 grant in February 2002 from the U.S. Trade and Development Agency. Of all things, this money was to be used to underwrite ITERA's effort to explore a Siberian gas field.[4] That a U.S. agency should agree to finance what in fact is a well-endowed Russian company, underwriting its efforts to explore for gas in Russia, does seem odd. The grant to ITERA bears a striking resemblance to the loan guarantees the German government gave to Nord Stream before Gerhard Schroeder lost his post as chancellor and became that company's chairman.

Although Weldon denies he did anything wrong, the circumstantial evidence that he was unduly supportive of ITERA is hard to ignore. Among other efforts on ITERA's behalf, Weldon sponsored a dinner in September 2002 to honor Igor Makarov, chairman of ITERA, at the Library of Congress in Washington; gave a speech in the House of Representatives in 2002 about ITERA; and participated in the opening of ITERA's headquarters in Jacksonville in 2003, a city a bit outside his Philadelphia congressional district.[5] The Russian paper *Kommersant* claims that between 2002 and 2004, Weldon also worked on behalf of two other Russian companies.[6] Nor did it help that all of this became public knowledge shortly after the lobbyist Jack Abramoff and disgraced Congressman Tom DeLay had also been accused of taking money from Naftasib, a Russian oil company, and then lobbying on its behalf.[7] For this, Abramoff was alleged to have been paid $2.1 million. With Russia's energy wealth, it was hard to avoid the perception that like many U.S. interest groups, a few Russian companies had also discovered how receptive the Republican-controlled Congress had become to financial incentives in 2006.

In October 2006, U.S. government attorneys obtained search warrants and then raided four houses and offices in the Philadelphia area and the ITERA office in Jacksonville. Weldon's efforts on behalf of ITERA then became a major issue in the election that November. Among the gaffes that became public was Weldon's praise for ITERA when he spoke at the opening of the company's headquarters in January 2003: "I can think of no other company that represents what Russia is today and offers in the future."[8] Given the accusations at home that ITERA was a classic case of Russian officials stripping assets from state companies for their own personal benefit, Weldon was probably more accurate than he intended to be.[9]

The Schroeder, Weldon, DeLay, and Evans seductions highlight not only how money can seduce some of the West's highest ranking officials but also how the Russians are learning to use their oil wealth outside Russia (they long ago learned how to use it within Russia). This wealth has catapulted Russia into new power relationships with its customers and Russia's main rival, the United States. Unlike the Cold War era, however, when the Soviet Union was effectively checked militarily by the United States and vice versa, it is hard today to find any similar restraint. Then, each feared to use nuclear weapons. Americans understood that if they used such weapons against the USSR, it would use its armed missiles against them. This was called MAD: mutually assured destruction.

NO MUTUALLY ASSURED RESTRAINT

Today, if the Russians or Gazprom threaten to halt the flow of their natural gas, there is little anyone can do about it. After twenty years or so, Russia's natural gas has become an integral part of the economies in the countries it serves. The European pipeline network does distribute gas from other countries but by far the greatest flow is from Russia. If the gas flowing from Russia—or the gas transiting from Central Asia in the Russian pipeline—were to be curtailed, consumers in Germany and other Central European countries near the Russian border would have a difficult time finding a substitute. While there are other producers of natural gas such as Algeria and Norway supplying product to the pipeline, they are at the other end of the pipeline network in the southwest or northeast, and it would be difficult to reverse the flow in a pipeline that has been designed to ship Russian gas coming from the east. Coal could be a replacement, but several months would be required to make the needed adjustments and the affected countries could experience several cold winters in the meantime. That is why we have likened the pipeline to an umbilical cord. Because Russia controls delivery of so much gas through its pipeline, in effect it has monopoly control in the markets it serves.

As we saw in Chapter 6, some have suggested that the Gas Exporting Country Forum (GECF) should turn itself into an OPEC counterpart. The group consists of fifteen countries, including the largest gas exporters, Russia, Iran, Algeria, and Qatar. As of 2007, GECF had met seven times, but despite some occasional tough talk, it has been unable to do much more than exchange information on such matters as technology, research, and perhaps some product swaps—for example, Algeria has sent its LNG (liquefied natural gas) to the United States in exchange for Russian natural gas delivered by pipeline to an Algerian customer in Western Europe.[10]

Some time in the future, the market for LNG may become large enough so that if gas delivered by pipeline should be cut off, LNG can be substituted in its place. But because producing LNG is so expensive and requires such a large capital investment, in 2005, 87 percent of it was sold only on the basis of long-term contracts. Unlike oil purchases that are made on the spot, LNG transactions often require as much as a two-year contract commitment. That in large part explains why as of 2005, 19 percent of the world's gas was delivered by international pipeline, compared to only about 6.9 percent as LNG.[11] LNG production is expected to double by 2010, but the likelihood is that relatively long-term contracts will still be required, making an actual spot market unlikely.

Even if some way can be found to produce LNG at a much cheaper price, remember that GECF members export only 14 percent of the gas consumed in the world. This compares to OPEC, whose members account for more than 35 percent of the world's oil exports. The United States, for example, produces 84 percent of the gas it consumes and Canada provides another 15 percent.[12] If Gazprom were to enter into a joint venture with BP in Trinidad to produce LNG, there might be local consequences for U.S. consumers accustomed to using imported LNG, but LNG constitutes such a small fraction of the total natural gas consumed in the United States that the Russian venture would have little impact on this country. The real threat of an effective GECF would be felt in Europe where Russia and Algeria account for 44 percent of Europe's natural gas consumption.

But there is reason to doubt that Russia would be willing to subordinate its actions to such an umbrella group. Russia has resisted membership in OPEC for that very reason. It prefers to let others coordinate production cutbacks. This allows Russia to increase its own production so it can benefit from the higher prices that OPEC cutbacks have created. It did just that in 1973. It is hard to see, therefore, why Russia would be willing to subordinate itself to a more powerful GECF. After all, it already has enormous clout.

This is why with Gazprom's newly acquired ability to determine economic, political, and personal success or failure, Russia is in a stronger position relative to Western Europe than it has ever been in its history. The threat of mutually assured destruction may have ensured no one would use missiles during the Cold War, but today there is no mutually assured restraint (MAR) to temper what might be called a one-nation OGEC, Organization of Gas Exporting Country, which is Russia today.

This is not to argue that Russia can do whatever it wants with its natural gas reserves. There are several concerns: one set has to do with the demand for Russian energy, one set has to do with its energy supplies, and one set is an offshoot of Russia's political and economic environment.

What would happen, for example, if there were a world recession and/or demand for Russia's gas and oil should suddenly slacken? After all, it is not as if Russia had no petroleum or gas reserves prior to 2000. Russia has been the world's leading producer of natural gas for some time, as well as a major, if not the largest, petroleum producer for many years. So why is it only now that Russia and Gazprom have become such concerns to the Europeans? The answer, in large part, is that as China and India have come of age financially, they have gobbled up most of the slack in the energy market. Despite the increasing emphasis on energy conservation, overall world demand continues to grow. At the same time, some of the existing reserves in Western Europe are being depleted. A far-reaching recession would undoubtedly precipitate a drop in commodity prices just as it did in 1997 and 1998, but at best that would be a temporary phase.

THE QUEST IN THE WEST FOR ALTERNATIVES

Compounding the problem, there are fewer and fewer prospects of finding new giant petroleum or gas fields. As in the past, the high energy prices of the early twenty-first century have stimulated the search not only for reserves that previously would have been unprofitable to exploit but also for new types of energy. Sources of supply such as the oil sands of Canada that until recently would have been too expensive to work now are worth developing. They provide Canada with reserves that some claim to be second only to Saudi Arabia's. Then there is also ethanol made from both sugarcane in Brazil and corn in the United States as well as solar and wind power. Detroit automakers have been campaigning to design cars that can run on either regular gasoline or E85, a combination of 85 percent ethanol and 15 percent petroleum.[13] They have already manufactured several million such flex-fuel vehicles as well as hybrids that utilize both electricity and gasoline. Similarly in the aftermath of the $90–$100 a barrel oil price, many in Europe as well as the United States and Asia are taking a second look at nuclear energy. Nuclear energy already generates almost 80 percent of France's electricity. However, it is unlikely that enough new nuclear facilities can or will be built in the near future or enough new forms of energy or enough new deposits of crude oil and natural gas in Russia or elsewhere can be found to replace the reserves depleted by existing consumption, much less to provide for an increase.

Additionally, Russia's influence could be weakened if the Europeans can find some way to gain access to natural gas from Central Asia without having to pipe it through Russia. At times, Kazakhstan has offered its support for various arrangements that would bypass Russia. (Almost as often it has disavowed any such routing out of loyalty to, and probably intimidation from, Russia).[14] This alternate route has at times been supported by the United States and the European Union. It should be a major priority for both Europeans and Americans.

Conceivably, LNG from non-Russian sources might also become a viable alternative, but at the moment, few places can supply enough of it at reasonable prices. Unless new technology offers a cheaper way of processing LNG, the prospects for creating a widely used spot market for LNG where consumers can purchase LNG at the last minute are not very good. For the time being, creating processing and handling facilities for producing, shipping, and distributing LNG is very expensive. As a result, almost everyone involved insists on a long-term contract commitment before they will agree to the necessary investments.

Of course, with time, there is the danger that other countries and other suppliers might try to encroach on "Russian markets." Alert to such a possibility, Gazprom, especially when Putin was president, could be counted on to attempt to gain control of possible producing fields outside Russia before potential competitors had a chance to intervene. For example, just as the Chinese have looked to Africa for sources of supply, so in 2008 Gazprom also sought to tie up what were thought to be very large gas deposits in Nigeria. This was seen as a strategy to gain control before private Western companies could step in, convert the gas to LNG, and use it to undermine Russian sales to Europe and the United States.[15] As part of the arrangement, Gazprom also promised to help Nigeria reduce the amount of associated gas released during oil production that the Nigerians flare, that is, burn off into the atmosphere.

But it is not only a question of whether or not there will continue to be such a strong demand for Russian gas. Several skeptics have also warned that Russia has overcommitted itself and has not invested enough in the development of new fields within Russia.[16] Among others, German Gref, formerly the minister of economic development, has complained that Gazprom has not only failed to expand productive capacity and maintain its existing infrastructure but it has also neglected commitments to re-equip and expand gas pipelines and other essential facilities, 30 percent of which he says needs replacement.[17] Furthermore, while neglecting essential producing facilities, Gazprom, he complained, has squandered capital on frivolous pursuits such as TV stations and newspapers. A report in the November 9, 2007, *Financial Times* warns that Gazprom is now spending more in expanding into other sectors of the economy than on developing new fields.[18]

A recent example is Gazprom's commitment to allocate $375 million to build a ski resort with three hotels, a covered parking lot for 1,000 cars, and a ski lift for the 2014 Sochi Winter Olympics. That comes with being a national champion. (Vladimir Potanin with his Norilsk Nickel and Oleg Deripaska with his aluminum company, Rusal, both national champions, are also diverting similarly large sums to Sochi.) But in the case of Gazprom, it is money that will not be going to the development of new gas reserves.[19]

Leslie Dienes of Kansas University has also pointed out that as long as domestic energy prices for both petroleum and gas in Russia are prevented from reaching market levels, those below-market prices not only subsidize excess consumption but they also discourage investment in the development of new reserves.[20] Prices are kept low for fear of a political backlash from the public if this benefit is eliminated. Dienes also points out that not only does the subsidized price for natural gas result in the misallocation of resources, but for the same political reasons, electricity rates are similarly controlled. Since natural gas is used to fuel almost half of the country's electrical generators, limiting electricity rates means that the electrical industry, along with the public at large, also has a strong interest in preventing any increase in natural gas prices.

Indicative of the problem, the Russian electrical industry has estimated that it will consume 186 billion cubic meters of gas by 2010. However, Gazprom predicts that electricity generation will need only 168 billion cubic meters, a difference and possible shortfall of 18 billion cubic meters.[21]

In an effort to use market prices to restrain demand and increase supply, the government has decided to "liberalize" domestic gas prices—

at least those that are paid by industrial users. To force nonindustrial consumers to pay much more was still deemed too risky. Nonetheless, by 2011, it is estimated that prices for industrial users will be double those of 2006.[22] Jonathan Stern, director of natural gas research at the Oxford Institute of Energy Studies, argues that if Russia increased natural gas prices as much as they were increased in Ukraine and Belarus, less would be consumed and there would be no problem supplying both the domestic and foreign markets.[23] John Grace agrees in principle but expresses it somewhat differently. "If domestic gas prices were at parity with the European market, there would be enough to supply both [markets]. . . . Even so it must be done at a very measured pace."

After widespread criticism that Gazprom had not invested enough to guarantee future production, on May 31, 2007, the government released a draft investment program spelling out what was needed to "Develop a Unified System of Gas Production, Transportation and Supply in East Siberia and the Far East." While the details made public are a little sketchy, this seemed to be an updated version of an energy development plan (whose details were also sketchy) covering both petroleum and natural gas that was first issued by the Putin government in 2003. If implemented, this latest version should forestall possible shortfalls. According to Alexander Ananenkov, acting CEO of Gazprom, by 2020 Gazprom should produce 670 billion cubic meters, a 14 percent increase in production.[24]

If production depended on Gazprom's investing enough in the development of future reserves, there could well be a shortfall in supply. But there are other possible sources of supply: independent gas producers in Russia, of which there are at least two, and gas produced as a byproduct by the country's several petroleum producers. In addition, Gazprom has also been able to count on reselling substantial quantities of gas from the Central Asian producers.

Admittedly, the gas production of Novatek and ITERA, both of which are independent of Gazprom (at least officially; in fact, Gazprom owns almost 20 percent of Novatek's stock), amounted to only 40 billion cubic meters. Byproduct gas produced by the country's petroleum producers equals another 40 billion cubic meters or so, which when added to that produced by Novatek and ITERA, equals about 15 percent of what Gazprom produces, which is a substantial amount. However, unless Gazprom allows Novatek and ITERA access to Gazprom's pipeline, these non-Gazprom producers are limited to supplying consumers within a short radius of their operations. Similarly, in most cases the petroleum companies can produce more natural gas but refuse to do so, because if they want to sell it to customers farther afield or in foreign countries, they too have to gain access to Gazprom's monopolized pipeline and Gazprom typically refuses access to any gas not produced by Gazprom itself. This exclusion is a legacy from the Soviet period.

On the rare occasion in the post-Soviet era when Gazprom has agreed to buy gas from a petroleum-oriented company like Yuganskneftegaz, it paid almost nothing for it. For example, in 2006, Gazprom paid less than $11 per 1,000 cubic meters for gas it grudgingly agreed to buy from Yuganskneftegaz.[25] That contrasted sharply with the $100 per 1,000 cubic meters that even Belarus was paying at the time.[26]

Since access to Gazprom's pipeline is not always possible nor profitable, the easiest way for the petroleum companies to dispose of their byproduct gas (associated gas)—and what for them is often a bothersome nuisance—is to flare it. According to an estimate by the French energy specialist, Pierre Terzian, Russian companies flare anywhere from 15 to 16 billion cubic meters of gas, an enormous waste. In fact a study prepared for the World Bank concludes that Russia flared more gas than any other country, twice as much as Nigeria, which was second only to Russia.[27] This also contributes needlessly to earth warming.[28] In part this was also due to the fact that until 2005, when Gazprom moved to absorb Sibneft, Gazprom did not seriously concern itself with petroleum production. It was almost completely absorbed in producing natural gas, so it had no interest in utilizing the associated gas produced as a byproduct by the country's petroleum producers.

When the petroleum companies spun off from Rosneft became privatized, their quest for profit led them to sell both gas and petroleum for the first time. Yet with a few exceptions, they profited little from their natural gas production because Gazprom continued to deny them access to the Gazprom monopoly pipeline. The attitude of what the Russians refer to as Gazoviki was "You, the petroleum producing companies, have no business impinging on our natural gas production activities."[29]

For those addicted to conspiracy theories, there seemed to be another reason for Gazprom's refusal to allow pipeline access. In several cases, refusal to allow some of the petroleum companies access to the gas pipeline was a part of the continuing effort by Putin and his Kremlin associates to regain control of properties given away during the privatization era. Gazprom's refusal to allow petroleum producers access to its pipeline was a way of preventing nonstate oil producers from following their contractual commitments to deliver gas and thereby forcing them to return ownership of potential gas fields to the country's national champions. In other words, this was just another form of renationalization designed to look like an initiative undertaken by the private producer, not the state.

GAZPROM: A STATE UNTO ITSELF

As an example, Gazprom refused to allow TNK-BP to build a pipeline so it could transport the gas it was producing in its Kovykta field in East Siberia to either large domestic or foreign markets. This was important because according to the terms of their license, TNK-BP promised that by April 2007 they would be producing 9 billion cubic meters of gas a year.[30] But as the time approached and Gazprom refused pipeline access, TNK-BP found that they could find a market for their gas only in Zhigalovo, a nearby logging town, and only for slightly less than 1 billion cubic meters. This was despite the fact that TNK-BP was charging a bargain $30 per 1,000 cubic meters when Gazprom was exporting gas at an average of $190 per 1,000 cubic meters.[31] Of course, transportation costs and market conditions in Western Europe and East Siberia are not the same, so there should be some difference in price, but this still seems to be inordinate. Beyond that, the only way TNK-BP could reach other customers would be for Gazprom to let them build their own pipeline network to the border. Potentially, Gazprom could build a pipeline network that could serve larger foreign customers throughout Asia. TNK-BP even offered to build their own pipeline for that purpose but were denied permission to do so.

Since TNK-BP could not reach any large customers if they were to produce gas, they would have had to burn it off—not only a waste but against the law and harmful for the atmosphere.[32] Alleged violations of other production commitments and charges of pollution forced Royal Dutch Shell and their Japanese partners to sell more than half of their holdings in Sakhalin II to Gazprom. It did not do anything for Russia's reputation when Shell and the Japanese agreed to sell Gazprom a half interest (plus one share of stock) for $7.45 billion. Most estimates set the value considerably higher.

While such wasteful and narrow-minded behavior has benefited Gazprom economically in the past, if it should be unable to fulfill contracts with gas from its own fields, there is the possibility that it can always seek to supplement its deliveries with gas from the petroleum companies. This may result in only a relatively small percentage of what Gazprom delivers, but we don't know. Until now, because of Gazprom's determination to maintain its monopoly, independent gas producers as well as the petroleum companies have been pressured to curb, not expand, gas production. If instead they were to be rewarded for increasing their gas output, output would undoubtedly be considerably higher than it is now.

There have already been hints of a change in policy. After the government agreed to raise the price for domestic industrial users of gas, Gazprom assumed that this would stimulate so much production by others that by

2020 Gazprom would be producing only 65 percent of the country's gas, considerably less than the 87 percent it produced in 2004.[33]

Gazprom's acquisition of the petroleum producer Sibneft has also led to a change in Gazprom's thinking. Now that a Gazprom subsidiary, renamed Gazprom Neft, is also drilling for petroleum, Gazprom finds itself producing the associated gas that comes up with the crude oil from the well. Gazprom now realizes that flaring much of that byproduct is a lost profit opportunity. Consequently, in August 2007, Gazprom announced that it had set itself the goal of utilizing 95 percent of what is called the "associated petroleum gas" (APG) by 2012. They extracted and used 14 billion cubic meters of APG in 2006 but expect to increase that to 22 billion cubic meters by 2011.[34]

Along the same lines, there is a growing governmental awareness of how much potential wealth is simply going up in flames. Even Putin has expressed his concern. In August 2007, he held a meeting with the heads of Transneft, Rosneft, and Gazprom, his favorite national champions, and warned them that if they burn off more than 5 percent of the APG they release into the atmosphere they will be fined. According to an estimate of the Ministry of Natural Resources, that could mean that the country's petroleum companies (both state and private) will have to pay $580 million a year because, according to Putin, Russian oil companies burn off more than 20 billion cubic meters of APG a year.[35]

Gazprom has been able to count on one other source of supplemental natural gas. Until they can find some alternative routing, Central Asian producers such as Turkmenistan, Kazakhstan, and Uzbekistan will have to continue shipping their gas to Europe through Russia via Gazprom's pipeline. As we saw when Gazprom cut off the flow of gas to Ukraine, much of that gas actually had been coming from Turkmenistan. Undoubtedly, the Central Asian producers will in time find some way to reach European markets by bypassing Gazprom pipelines, but for the near future most of their exports will continue to flow through Russia and at least some of this gas will continue to be resold by Gazprom as part of Gazprom's contractual commitments.

Another alternative is for the Central Asian countries to turn east or south and sell to China, Iran, or India. Kazakhstan has done just that with its oil pipeline to China. The Kazakhs have also indicated their support for Turkmenistan, which has signed an agreement with China to build a new natural gas pipeline that begins in Turkmenistan and crosses Kazakhstan and Uzbekistan. It would supply 30 billion cubic meters of natural gas a year for thirty years. Access to gas from any of the three Central Asian suppliers would also allow China to bargain harder with Russia and Gazprom over prices. The Chinese are notorious for their hard bargaining over prices, and negotiations with the Russians have frequently broken down when no agreement could be reached.[36] This has not contributed to better relations between the two countries. In 2007, the Chinese refused at first to agree to a price of $100 per 1,000 cubic meters. They reluctantly agreed to go up that high but the Russians wanted at least $125 per thousand cubic meters, pointing out that as of 2011, that would be the price within Russia itself. Even then this would be but half of what the Europeans would be paying.

WHAT ABOUT TOMORROW?

Whether or not the Russians will have enough natural gas to meet their contracts, some Russians for the first time are beginning to question the wisdom of seeking a continuing increase in petroleum and gas output. What is wrong, they ask, about stabilizing output or even discouraging or reducing production? By contrast, those seeking to increase output in the petroleum industry often criticize the tax on petroleum exports that claims most of the revenue collected once the price exceeds $27 a barrel. But proponents of a freeze in production support such a tax because it discourages, temporarily at least, investment in further exploration and development. This postpones such exploration for now and means there will be more set aside for the future.

Yet if production of either natural gas or petroleum should drop or even stabilize, Russia may be unable to meet all of its export commitments. In some cases this might result in a contract violation (something not unprecedented in Russia). But since the rule of law involves a less than fully formed yardstick at best, curbing output and then failing to honor agreed-to contracts would not be regarded by many Russians as a particularly serious crime.

While he did not support any effort to break existing contracts, Sergei Karaganov, deputy director of the Institute of Europe, wondered aloud at a meeting of the Valdai Hills group in Moscow in September 2006 why Russia was so intent on increasing its annual production of petroleum and gas. Rejecting arguments that Russia should take advantage of the moment when prices were at a record high, he asserted that Russia had already collected more revenue from foreign energy sales than it knew how to spend. Look at the $100 billion put aside in the Stabilization Fund as of September 2006 when he spoke, not to mention the more than $300 billion hoard of foreign currencies and gold accumulated in the Russian Central Bank vault. Besides, if their energy reserves were valuable in 2006, given the growing world demand it was more than likely that such deposits would be even more valuable in the future. Further, he noted, the demand for natural gas is fairly inelastic, at least among already existing purchasers. Thus if supplies were reduced, the price for the gas that is being sold would most likely rise so that total revenue might actually increase.

President Putin, however, takes a different stance. At the 2007 Valdai Hills meeting he insisted that Russia should produce as much as it can immediately before someone discovers an energy substitute that leads to a drop in oil and gas prices.

JUST THE FACTS, PLEASE

Thus while some insist that Russia has not invested enough in new field development and production to assure that it can honor its existing contracts, others argue that Russia should sell its oil and gas as quickly as possible rather than set aside reserves for the future. But what has actually been happening?[37] Data taken from BP's annual statistical survey and used in Table 7.2 suggest that despite warnings that growing consumption within Russia will soon make it impossible to set aside enough oil and gas to meet export obligations, so far Russia seems in no such danger, at least in the short run. In the case of petroleum for example, as of 2006, the amount produced and available above and beyond domestic needs has, if anything, increased each year.

As for natural gas, the trend is not as pronounced. The amount available in 2006 was 10 percent less than it had been in 2005, certainly a worrisome sign. Yet the amount available has fluctuated up and down from year to year. Thus while the export potential was also less in 2006 than in both 1996 and 1999, the amount available for export in 2006 actually was higher than it was in 2000, 2001, and 2002. Russian authorities would do well to discourage wasteful consumption and encourage investment in new fields, but, for the time being at least, there appears to be a comfortable cushion.

FIGHTING OVER THE SPOILS

Yet by no means is everything in the Russian energy sector is going well. Below the surface, major battles are being fought. When there is so much wealth up for grabs, there are bound to be major battles over who should control it. Certainly Russia is not unique in this respect, but since the state is so deeply

TABLE 7.2 Russian Natural Gas and Petroleum Available for Possible Export

Natural Gas (in Billions of Cubic Meters)

	1996	1999	2000	2001	2002	2003	2004	2005	2006
Production	561	551	545	542	555	579	591	598	612
Domestic Consumption	380	364	377	373	389	393	402	405	432
Available	181	187	168	169	166	186	189	193	180

Petroleum (in Millions of Metric Tons)

	1996	1999	2000	2001	2002	2003	2004	2005	2006
Production	303	305	323	348	380	421	459	470	481
Domestic Consumption	130	126	124	122	124	123	124	123	129
Available	173	179	199	226	266	298	335	347	352

Source: *British Petroleum Review of World Energy, June 2007, pp. 8–15, 24–29*

involved and since the struggle for control is far from settled, the fight in Russia has some unique features—what I have called "The Russian Disease."[38]

Unlike the Dutch Disease, which, as we noted in the Introduction, is shorthand for the negative impact on a country's manufacturing competitiveness that is a consequence of a major discovery of natural gas or oil, the Russian Disease has more to do with the greed and the jockeying for control and ownership that hit Russia during the privatization process. It was a particularly acute problem because Russia lacked what the Wellesley College economist Karl Case has called a "moral infrastructure" where there are few of the firmly rooted commercial laws or informal moral codes that are taken for granted in long-established market economies. The "rule of law" becomes the "the law of rulers." As a result, when a country like Russia suddenly decides to privatize its valuable energy assets, the law of the jungle will almost inevitably take over and lawlessness and chaos will ensue. A blatant instance of this occurred when Putin formally announced that Gazprom would take over state-owned Rosneft. In what later turned out to be an embarrassing publicly televised announcement, on March 2, 2005, Alexei Miller, the CEO of Gazprom, appeared together at Gazprom headquarters with Sergei Bogdanchikov, the CEO of Rosneft. They used the occasion to congratulate each other on reaching an agreement to merge Rosneft and Gazprom under Gazprom leadership.[39] Yuganskneftegaz, Rosneft's largest producing unit, had just been seized from Yukos, and the announcement was that it would become a separate state-controlled entity. The rest of Rosneft was to be incorporated as part of Gazprom. With Bogdanchikov at his side, Miller confirmed that "a final decision on the procedure to join [unite] Rosneft with Gazprom has been made and the state will receive a controlling stake [more than 50%] in Gazprom."[40] In a bow to Putin's determination to create national champions, Miller added, "Integration of the assets of Gazprom and Rosneft strengthens Gazprom's presence in the oil sector [where it previously had almost no production] and allows the company to become, in the very near future, one of the biggest gas, oil, and energy companies in the world."

Bogdanchikov, however, evidently had some reservations. While he sat smiling throughout the TV presentation, he pointed out, in a somewhat off message aside, that he would head up Yuganskneftegaz as a separate entity outside of Gazprom. This seemed to contradict Miller, who insisted that the "de facto consolidation" of the two companies would take place "in the near future" and that Yuganskneftegaz would be incorporated as part of Gazprom. This had been the original intention when Yukos was seized by the government in December 2004.

The reason Yuganskneftegaz was not immediately absorbed into Gazprom was concern that disgruntled Yukos stockholders would soon launch a lawsuit in an attempt to seize some of the many Gazprom assets located outside of Russia. As soon as the likelihood of such a Yukos stockholder lawsuit faded, Yuganskneftegaz would then also become part of Gazprom. This would be another step toward enhancing Gazprom's role as a national champion. But with or without Yuganskneftegaz, it seemed clear that Miller and Gazprom would soon assert control over Rosneft and perhaps eventually Yuganskneftegaz.

To the shock of many observers, however, the very next day Bogdanchikov insisted that no, he had not agreed to merge Rosneft into Gazprom. In fact, a press release issued by Rosneft flatly contradicted Miller. "These statements [about the merger] do not reflect reality and they seem to be seen exclusively as the Gazprom CEO's personal opinion." (But what did we hear the day before on television?) In response, Gazprom's press office described the Rosneft assertion as "a technical mistake"—in other words, a lie! Continuing the soap opera, the Kremlin then issued a statement announcing that Rosneft had retracted its statement, which in turn provoked Rosneft to deny it had done any such thing.[41] This event provided an unprecedented look at the bureaucratic cat fight then going on within the Kremlin and evidence that even Putin was unable to impose a coherent policy. There were then allegations by Rosneft that in the original TV presentation, Bogdanchikov had in fact insisted that Gazprom and Rosneft would indeed remain two separate legal entities, but that the segment shown on TV omitted that portion of the announcement. The Rosneft spokesman pointedly explained that the TV station reporting the announcement was owned by Gazprom and had deliberately sought to mislead the public. Shocking! Seeking to defend Rosneft, the spokesman went on to say, "Why this happened is a question for Miller." (Who is in charge here?)[42] In the end, Rosneft did remain a separate company and did hold on to Yuganskneftegaz, the prime cherry in the orchard.

The fact that Rosneft and Bogdanchikov, along with Igor Sechin, who sat both as chairman of Rosneft and deputy head of the Kremlin administration, dared to defy Gazprom and Miller reflects how determined the senior executives of Rosneft were to hold on to their priceless perquisites. This case study in the Russian Disease displayed a remarkable show of self-confidence by Rosneft senior executives. It was a display as bold and disrespectful as anything Khodorkovksy ever did and suggestive of the fight over the spoils that seemed to be taking place once it looked as if Putin would soon be leaving the office of president.

The difference in response to such insubordination is due to the fact that unlike Yukos and the outsider Khodorkovksy, Rosneft is controlled by a special group of insiders, what the Russians have come to call "siloviki." There is no exact equivalent for the phenomenon in the United States and Europe, and so no word for it exists in English. The Russian word "sila" means strength, so perhaps the best way to convey the concept is to translate it as "law and order veterans." Russians use the word to refer to former members of the KGB and to a lesser extent senior military officers and police. Given Putin's own background as a lieutenant colonel in the KGB, it is not surprising that he has brought such people into his government.

Olga Kryshtanovskaia, a sociologist who specializes in the backgrounds of senior government appointees, has compared the number of siloviki in the senior ranks of the Gorbachev, Yeltsin, and Putin governments.[43] She found that the number of siloviki in the national leadership rose from about 5 percent in senior positions under Gorbachev to 58 percent under Putin.

Balancing off these law and order types, Putin has also brought in a substantial number of technocrats, former colleagues who worked with him in Mayor Anatoly Sobchak's St. Petersburg office, where Putin headed the Department of Foreign Economic Relations. This included Viktor Zubkov, Sergei Ivanov, Viktor Ivanov, Dmitry Medvedev, Igor Sechin, Sergei Naryshkin, Minister of Finance Alexei Kudrin, and Minister of the Economy German Gref (see Table 7.3). These FOP (Friends of Putin), called by others the St. Petersburg "mafia," are similar to the Texas "mafia" brought to Washington by the U.S. President George W. Bush and the Arkansas mafia that came in with President Bill Clinton.

So ubiquitous are these outsiders from St. Petersburg that they have become a source of humor among Moscovites jealous of these officials from the provinces who have taken over their city. Resentful Moscovites tell the story of a local man riding the Moscow subway. Without warning, the man next to him steps on his foot. After five minutes, the Moscovite, in pain, works up his courage. "Excuse me, sir," he says. "Are you in the KGB?" "No," answers his neighbor. "Then are you from St. Petersburg?" "No," again comes the answer. "Then tell me, why are you standing on my toes?"

TABLE 7.3 Siloviki in Business

Name	Title	Business	Concentration	Day Job
Sergei Chemezov*	Chairman / CEO	Rosoboronexport	Arms exports	
Anatoly Isaikin	CEO	Rostekhnologi	High tech manufactured goods	
Sergei Ivanov*^	Chairman	United Aviation	Airplane manufacture	1st Deputy Prime Minister
Viktor Ivanov*^	Chairman of Board	Aeroflot; Almaz-Artey	Airline; Air defense systems	Deputy head Kremlin administion
Igor Levitin	Chairman	Sheremetyevo Airport	Airport	Minister of Transportation
Dmitry Medvedev^	Chairman	Gazprom	Natural gas	Former Putin Chief of Staff; First Deputy Prime Minister
Sergei Naryshkin*^	Vice chairman	Rosneft	Oil	Deputy Prime Minister
Sergei Prikhodko	Chairman	Tvel	Nuclear fuel trading	Foreign affairs adviser to Putin
Igor Sechin*^	Chairman	Rosneft	Oil	Kremlin staff
Anatoly Serdyukov	Chairman	Khimprom; Rostekhnologi	Chemicals; High tech manufactured goods	Minister of Defense
Yevgeny Shkolov	Board of Directors	Transneft	Oil pipeline	Presidential aide
Igor Shuvalov	Board of Directors Chairman	Russia Railways; Sovcomflot	Railway; Shipping Company	Presidential economic adviser
Sergei Sobyanin	Chairman	Tvel	Producer of nuclear fuel	Putin Chief of Staff
Vladislav Surkov	Chairman	Transnefteprodukt	Pipeline hardware	Kremlin staff
Vladimir Yakunin*	President	Russia Railways Co.		

* former member of KGB
^ from St. Petersburg

But the siloviki around Putin and the privileges and power they have been given distinguish them from their Washington counterparts. The Washington political appointees are very often appointed to plum positions *after* they leave government service. Under Putin, they are appointed to the lucrative posts *while* they hold their jobs in the government. Given the way that new government officials in Russia treat their predecessors (a legacy of the USSR when predecessors were blamed for all the country's subsequent problems—Brezhnev, Gorbachev, Yeltsin, and Stalin are good examples, not to mention Khrushchev, who was put under house arrest), Russian officials fear that once they are out of the government, they will no longer have access to such patronage. (This habit is also a source of humor. As he is about to turn over his office to his successor Vladimir Putin, Boris Yeltsin is asked by Putin if he has any words of advice. Yeltsin tells Putin that if he encounters any difficulties, he should look in his desk in the president's office where Yeltsin will leave three envelopes. At the first sign of trouble, open the first envelope. Sure enough, Putin has trouble and decides to open the first letter. It reads: Tell everyone it was [Yeltsin's] my fault. If the trouble continues—open the second envelope. It reads: Promise that everything will get better. If the protests still continue, prepare three envelopes.) So Russian officials have learned to take such posts and collect the fringes they bring while they can.

DOUBLE DIPPING

Whatever the rationale, Putin has instituted the practice of appointing his former colleagues not only to the most senior positions in the government but also to senior and lucrative posts in the business world. One Russian told me that senior Gazprom executives are reported to collect salaries that exceed $300,000 a month. (That surely would be high by past Russian standards but still less than what some senior executives in the United States receive today.) But unlike the practice in the West, the Russians hold these appointments simultaneously with their government positions. Since many of the posts are at the top of Russia's most prosperous companies, they are following in the steps of the oligarchs, only they, unlike Khodorkovsky, Berezovsky, and Gusinsky, are protected by their benefactor, Putin. These FOP are doing very well. The *Moscow Times* on June 22, 2007, for example, reported that several of them have joined some of the old oligarchs in paying the $50,000 a year dues to the Burevestnik Yacht Club, Russia's largest, which even has its own helipad. The yacht is extra.

This was not the only glimpse into what seems to be the acquisition of great wealth by KGB veterans or what we have called the second generation of oligarchs. In an interview in the Russian paper *Kommersant*, Oleg S. Svartsman acknowledged that the $3.6 billion Finansgroup investment fund that he manages is a vehicle used by members of the FSB and SVR (Foreign Intelligence Service) and other high-ranking officials and their families. They use this to enhance their wealth (at $3.6 billion, this is pretty good for a relatively small group of public servants) and promote "social responsibility to the state."[44] In the words of Yevgenia Albats, the deputy editor of *New Times* and the author of *The State Within a State*, a book about the KGB, "The FSB is no longer just a police organization, it is a business."[45]

As a look at Table 7.3 shows, Putin appointees now run most of Russia's national champions. At a meeting of the Valdai Hills group on September 9, 2006, at Putin's home in Novo-Ogaryovo outside of Moscow, I asked President Putin if this was a good way to run both the government and these companies. "How could someone like Igor Sechin," I asked, "the deputy head of the Kremlin administration, which is a full-time job, also do a good job as the chairman of the board of Rosneft, which is also a full-time job? More than that, how could Sechin be expected to be objective if the Kremlin is asked to mediate a dispute between Sechin's Rosneft and an ostensibly private company like Lukoil? Wouldn't this constitute a clear conflict of interest and open the door to criticism of corruption and abuse of power? Moreover, aren't they likely to enrich themselves in the process, and before long end up as new entries on the *Forbes Magazine* list of world billionaires?"

Putin's response, according to the official transcript, seemed to skirt or, perhaps more charitably, miss the point. "These people only represent the state's interests in a given company where the state holds a certain number of shares. . . . They do not manage the company and do not manage its resources. . . . Of course we could develop a system in which independent experts and lawyers represent the state's interest. . . . For now I consider this is not realistic because at present lawyers and independent managers would at once start to engage in their own private business." He then cited as an example of such corruption the two Harvard University advisers sent to Russia to help with the privatization effort only to be sued, found guilty, and fined $2 million each by the U.S. government for violating their work contract and insider trading. "For that reason it is natural that bureaucrats working for the state should represent the state's interests."[46]

Whatever the reasoning, it is hard to find any other country in the world where so many senior government officials can simultaneously hold such lucrative business positions, whether in wholly or partially owned government companies. It is also hard to see how such companies can avoid patronage and other political pressures—pressures that are often difficult for even purely private companies to ward off. Inevitably, the partially private Russian companies headed by siloviki or FOP face the same political pressures encountered by wholly state-owned companies over such issues as patronage appointments, where to locate factories, pricing, wage setting, and vendor selection. In most such instances, production, productivity and profits suffer.

It may already be too late. Russian state companies have begun to appoint the children of siloviki to high corporate positions. Called "princelings" in China, their Russian equivalents are a form of a kickback for senior leaders in the party and government. While these Russian princelings may be capable, it is clear to other employees that their appointment was due more to their parentage than their training or competence. As Table 7.4 shows, these appointments are to senior posts in state-owned entities. Nikolai Patrushev, head of the FSB (once the KGB), for example, has managed to place two sons in these companies.

At the 2007 Valdai Hills meeting I asked Putin if appointing the children of government officials, who would probably otherwise not be regarded as qualified, would likely have a negative impact on company productivity. Putin responded that such impact would be minor and besides

TABLE 7.4   Princelings

Parent	Child
Valentina Matviyenko Governor, St. Petersburg	**Sergei** Senior VP, Vneshtorg Bank
Sergei Ivanov Minister of Defense Deputy PM	**Sergei** VP, Gazprom Bank
Mikhail Fradkov PM	**Pyotr** Deputy General Director, Fesco (Far East Shipping Co.)
Nikolai Patrushev Director, FSB (FBI)	**Andrei** Advisor to CEO, Rosneft
	Dmitri Vneshtorg Bank

such practices are not unique to present-day Russia. They are common all over the world. To illustrate his point he related a joke from the Soviet era. An official is asked if it is possible for a Soviet general's son to become a general. "Sure, why not?" "Then is it possible for a general's son to become a marshal?" "No!" "Why not?" "The marshals have their own sons."

While nepotism of this sort may have been commonplace under the Czar, except for the generals and the marshals there was very little of it in the Soviet era. At best the children of Politburo members would sometimes find appointments in academic institutions, although Brezhnev's son-in-law had a senior appointment in the Ministry of the Interior. Other than a few such exceptions, when it comes to feathering the beds of the children of those close to Putin, Russia today more closely resembles China than the USSR. As such, there is a strong likelihood that productivity will be affected. For example, the Japanese economist T. Shiobara points out that it costs Gazprom $3.3 million to construct one kilometer of gas pipeline whereas the world average price is approximately $1 million.[47] When asked about such disparity, Putin pointed out that Russia builds many of its gas lines in permafrost and swampy conditions and so costs should be higher. "Russia is not the Balkans," is how he put it. Yet a threefold higher cost does seem excessive.

There are other signs of inefficiency and waste that are most likely a consequence of state ownership. John Grace has found this to be the case in the petroleum industry. As for gas production, Michael D. Cohen, an industry economist at the U.S. Department of Energy's Energy Information Administration, cites data from his office that in 2004, "roughly 70 billion cubic meters [or one-third of Russia's gas exports] either leaked in the form of methane in the course of transmission or distribution . . . or were flared," although some of the flaring was by private petroleum companies. Cohen also noted that only 20 percent of Gazprom's investment is directed to upstream or new projects that will result in increased production.[48]

On occasion, some Russians in the private sector have been bold enough, or perhaps indiscreet enough, to warn that the state was again becoming too involved in the economy, especially in the energy sector. Vagit Alekperov, the CEO of LUKoil, did just that in August 2007. Vladimir Bogdanov the CEO of Surgutneftegaz, said much the same thing a few months later in October. Alekperov later backtracked, but it is clear there is concern that if the state is allowed to become too dominant, it will affect efficiency throughout the whole economy for all the usual reasons associated with state ownership and management of economic resources.[49]

In much the same spirit, once anointed as a national champion, there is a real danger that a company like Gazprom may become a law unto itself and act as if it is exempt from the normal checks and balances that restrain ordinary corporations. Gazprom, for example, despite widespread protests, has decided to build itself a new skyscraper monstrosity in St. Petersburg at the mouth of the Okhta River across from the city center. What makes such a decision so controversial is that this new structure will be 396 meters high. Therefore, in this eighteenth-century city it will tower over the rest of the buildings, which are restricted to 48 meters.[50] Moreover, once one building is granted an exception to the height restriction, it is all but certain that others will also want the same privilege and before long, the beauty that makes St. Petersburg so unique will be destroyed. Some local environmental groups as well as UNESCO have protested, but there is not much likelihood that a company as strong financially and politically as Gazprom will be prevented from moving ahead with construction.

SOCIAL PROBLEMS

While deciding how to control these new corporate entities is bothersome, Putin and his successors are faced with other issues that may be even more intractable. At one time or another Putin has indicated his awareness of most of these problems. To deal with the continuing decline in the Russian population, which has been shrinking by almost 700,000 a year, he has proposed a bonus program for Russian women to encourage them to give birth to more than one child. Russia's population shrinkage makes it difficult to find enough young men to staff not only the army but also the industrial and agricultural workforce. As in several West European countries, a declining birth rate means that it will be harder and harder to support the increasing percentage of the population who have reached retirement age. In addition, Russia has to contend not only with a low birth rate but also with a life expectancy for men that hovers between fifty-eight and fifty-nine years. That is a disgrace for an industrial country. Excessive drinking, poor diet, automobile accidents, and violence among men account for much of the problem. As Putin pointed out in one of our Valdai Hills meetings, Russian men who live to age sixty-five have a life expectancy thereafter that is comparable to that in the West; the trick is to make it to that point.

Even if the population were growing, Russia would have a problem finding people to live in the Maritime Provinces of the Russian Far East. The population there is falling even faster than the national average. Many of those who remain want to move to Moscow and the center of the country. According to the Institute of Economic Research in the Far East, the population there has fallen 16.5 percent since 1989. As a consequence, only 4.6 percent of the total Russian population is left to occupy 36 percent of the country's land.[51] The depletion of the population there is of concern for economic as well as political reasons: while the Russian population is shrinking (only 6.6 million Russians resided there in 2006), the Chinese population immediately adjacent to the Chinese-Russian border along the Amur River is growing rapidly and totaled 38.1 million in 2006.[52]

When asked at the 2006 Valdai Hills meeting what he would have liked to have remedied but so far had not, Putin mentioned not only the shrinking population rate but also the national poverty level: large numbers of Russians live far below the poverty line despite the substantial improvements of recent years. He also expressed frustration over his inability to fight corruption. There are many indicators to suggest that corruption has become an even more serious problem on Putin's watch. This may explain why a year later he appointed Viktor Zubkov as prime minister; Zubkov had a reputation for successfully dealing with corruption.

Putin also expressed his hope that he might yet do more to develop the country's political system. Given that Putin has been criticized both inside and outside Russia for doing away with elections for governors, eliminating diversity of views in the media, and purging the Duma of any meaningful opposition to one-party rule, his reply suggests that he is either blind to his own shortcomings or his interpretation of freedom of the press and parliamentary democracy differs from that normally held in the West. It may be a little of both.

Mention should also be made of the continuing unhappiness with Russian rule in parts of the country, particularly in Chechnia and some of its neighbors in the Caucasus. Putin has managed to subdue much of the open anti-Russian resistance, but few would insist that the region has been fully pacified. The battle there, just as happened to the United States in Vietnam and again in Iraq, has had the effect of bloodying and discrediting the army, and it will take some time for those wounds to heal.

As talented as he has been in resolving so many of his country's problems, there seemed to be yet another difficulty that even Putin could not solve: what would happen to the political and social structure he has set up after he left office. Putin did a magnificent job in ending the political and economic free-for-all of the Yeltsin years. He established firm control throughout the government, removing most of the old oligarchs and installing FOP, both siloviki and economic liberals. But because so many of these KGB alumni have become new oligarchs with their own little and not so little economic empires, it was uncertain whether or not they would submit to a Putin successor to whom they owed nothing. Putin had the advantage of being able to pass out patronage plums and in doing so build up a group of loyalists, just as the former Czars did. Even under Putin, however, the prospect of controlling so much wealth has on occasion led to open feuding. A good example was the unseemly brawl between Miller of Gazprom and Bogdanchikov of Rosneft when the Rosneft group refused to submit to Gazprom control and effectively called the Gazprom executives liars.

Nor was this the only instance where rivalry among Kremlin insiders had surfaced in public. In 2007, Vladimir Kumarin, the owner of the St. Petersburg Fuel Company, a chain of gasoline filling stations, was arrested. According to the newspaper *Novaya Gazeta* of September 10, 2007, Kumarin's arrest was initiated by the Igor Sechin faction in the Kremlin against the anti-Sechin siloviki. In another case, the attempted arrest of Mikhael Gutseriyev and the takeover of his company, Russneft, as well as the struggle for control of the pharmaceutical distributor Protek and Biotek, also seemed to be a part of this internal battle for control of corporate assets and other valuable spoils that was being waged by this second echelon of siloviki and FOP oligarchs. There even seemed to be a struggle within the upper ranks of the FSB. For example, the arrest of Lieutenant General Aleksandr Bulbov, chief of one of the highest-ranking divisions of the Federal Narcotics Control Service (FNCS) was regarded as just such an instance of this "interagency warfare." This also seemed to involve a struggle between Nikolai Patrushev, the chief of the FSB, and his ally Igor Sechin, a KGB veteran and a deputy head of the Kremlin administration as well as chairman of Rosneft, on one side against Viktor Cherkesov, who is Bulbov's boss, on the other side. In this case the FSB was retaliating for an earlier raid by the FNCS against the FSB in 2000 and the subsequent dismissal in 2006 of a number of high-ranking FSB officials who were charged with the theft of valuable property and engaging in illegal real estate transactions. Undoubtedly some of this jockeying for control reflected a deep concern that Putin's successor would reallocate some of the country's assets and strip the new siloviki oligarchs of their assets just as Putin stripped the original oligarchs.

As if all this were not enough infighting to keep the gossips in Moscow busy, Sergei Storchak, a well-regarded deputy minister in the

Ministry of Finance, was arrested in late 2007 and charged with embezzlement of $43.5 million in state funds. Storchak had been in charge of administering the country's $100 billion-plus stabilization fund and a close and trusted ally of Alexei Kudrin, the minister of finance. Kudrin fiercely defended his friend Storchak, who he insisted was innocent.[53]

If even Putin could not control such open feuding, his successor was likely to have an even more difficult time herding these new aristocrats. Now that they not only have their old KGB know-how and connections but also large financial wherewithal, these new oligarchs will be much more difficult to control than the original oligarchs.

Concerns of this sort may account at least in part for Putin's decision to stay involved as a Russian leader. Putin was clearly aware that he would be damned if he stayed on in power (at least by the outside world). He explicitly said so in the September 2006 meeting of the Valdai Hills Discussion Group. As he put it, "I don't believe that the country's stability can be insured by one man alone. . . . If everyone is equal before the law, I cannot make an exception for myself . . . [and ignore the constitution that limits the president to two four-year terms]."

By 2007, however, with his popularity at 70–80 percent, he also came to understand that if he did not stay on, that would also be destabilizing. That led to speculation that Putin might agree to serve as prime minister or in some other vague czar-like role as "father of his country or national leader."[54]

Ultimately, Putin decided to announce his choice for president. This put an end to most of the speculation and jockeying to stake out a claim on state assets. His nominee was Dmitry Medvedev, who was not only the first deputy prime minister but also the chairman of the board of directors of Gazprom as well as Putin's longtime protégé and friend from their days in the St. Petersburg governor's office. Medvedev had followed Putin to Moscow and once there eventually became the head of the Kremlin administration. Regarded as a Putin loyalist, Medvedev certainly has had administrative experience, but his relative youth (he is only forty-two) and his lack of gravitas and experience with the KGB alumni made him vulnerable to sniping and intrigue from the siloviki.

Recognizing the danger, Medvedev (perhaps with Putin's own acquiescence and maybe even at his initiative) proposed that Putin take on the job of prime minister. Officially this seemed to represent a demotion, but it did leave Putin in place to protect Medvedev's flank from the siloviki, and it allowed Putin to retain influence and reassure the public without having to amend the constitution. (One way to judge who really is in charge in this new arrangement is to see if Putin and his family move out of their palatial presidential residence, where Putin welcomes groups such as the Valdai Hills Discussion Group, and surrenders it to President Medvedev.) Such an appointment is no guarantee that there will be no quarreling between Putin and Medvedev, but it did seem to put an end to much of the short-term uncertainty and instability. Given Medvedev's close involvement with Gazprom, it is also another indication of how important energy and Gazprom are to Russia's well-being.

ECONOMIC CHALLENGES

While Putin's selection of Medvedev may resolve some of the political issues, much also remains to be done to stimulate the non-oil, non-raw-material sectors of the economy. The prosperity that has accompanied the oil boom has been a key factor in improving Russian general economic health, but it has also harmed some sectors. The big boost in disposable income as a result of energy exports notwithstanding, the strong ruble has hurt Russian manufacturing efforts just as the strong guilder hurt manufacturing in the Netherlands. Except for vodka, caviar, and Kalashnikov weapons, almost no commercial products manufactured in Russia have ever won an international competitive preference. It is not that Russia had a strong competitive manufacturing sector before the jump in energy prices. Manufacturing in Russia both in the Czarist and Soviet eras has always needed government subsidies. According to *Izvestia*, today Russia's share of the world's high market is only .5 percent and its machinery exports total only .3 percent of world exports.[55] Steep tariffs or import protection barriers help domestic manufacturers compete against foreign imports, but the strong ruble complicates whatever efforts are made to foster domestic manufacturing. Because of tariff protection, a large number of foreign automobile manufacturers have opened assembly plants in Russia to take advantage of the increase in Russian consumers' disposable income. But if and when those tariff barriers are lowered as a condition for Russia's entry into the World Trade Organization, that is bound to hurt such efforts.

DO YOU WANT A RUSSIAN FOR A PARTNER?

With or without such tariff protection, many foreign companies will continue to search for opportunities to invest in Russia. Certainly the risks are high; remember the cases of General Motors, which was squeezed out of its joint ventures in Togliatti, and Dutch Shell, which was forced to bring in Gazprom as a partner at an enormous discount. Still, with its oil wealth and growing disposable income, Russia offers a rich market. But there is little prospect that foreign companies will be able to count on the rule of law to protect their property, especially when a national champion takes an interest in their activities. By the same token, Russian companies should not be surprised if they are treated unfairly when they seek to operate outside Russian territory. Many Russians, including Putin, were upset when the Russian steel manufacturer Severstal attempted to merge with Arcelor, the West European steel company. At the last minute, Arcelor opted to dump Severstal and merged instead with Mittal, an Indian company. Most Russians viewed this as a form of discrimination against Russia and Russians. There was a similar reaction when Russians were told that they would not be allowed to have a representative on the board of directors of EADS, the parent company of Airbus, even though Vneshtorgbank had purchased between 5 and 7 percent of the company's stock and for a time considered buying as much as 10 percent. In the same way, Aeroflot was rejected when it sought to take over operating control of the troubled Italian airliner, Alitalia. In 2006 alone, Epsilon Corporate Finance reported that Russian companies made twelve attempts to buy shares in European companies, all of which were rejected. Five of those efforts involved Gazprom—most notably, offers for Centrica in England and PGNIG in Poland.[56]

One reason for such discriminatory treatment is the real fear of the symbiosis between Russia's national champions and the state. If a Russian company is allowed to buy a share in a Western company, the Western company may find that its partner is not a commercially oriented investor but the Russian government—a corporation that in fact is a Russian national champion and President Vladimir Putin. That is certainly what some Western critics have concluded about the way Gazprom operates when it moves into Western markets.[57] This is an example, even if indirectly, of the price Russia must now pay for the disdain and disregard for the rule of law. Whether justified or not, Russian investments outside of Russia are now regarded with extra caution and will be treated skeptically for some time to come.

Certainly Russia's financial and economic condition has improved since the August 17, 1998, financial meltdown. As of fall 2007, all but a fraction of Russia's foreign debt has been paid and Russia now has a $420-plus billion reserve of hard currency as well as a $120 billion stabilization fund that had been administered by Sergei Storchak from the Ministry of Finance. But while the state has an impressive cushion of assets in the Central Bank, the corporate and local government sectors have taken advantage of eager and overgenerous Western investment banks to borrow sums that by some accounts will soon equal what the state has accumulated in its reserve rainy day fund. If there should be a serious drop in energy prices, the local governments that depend on the corporate sector taxes to pay their bills would face serious difficulties. The Russian companies that have taken advantage of Russia's high credit rating to borrow billions of dollars and euros would be in similar trouble. A really substantial collapse in energy prices would spell an end to Russia's status as a super energy power.

RUSSIA INVESTS ABROAD

But if energy prices do not drop, or at least do not drop significantly, the Russian government will almost certainly continue to seek to exert its influence outside of Russia for some time to come. We can also expect to see a larger and larger presence of Russian companies outside of Russia. Rather than just deposit U.S. dollars in a bank or invest them in purchasing U.S. government securities, both the Russian government and Russian individuals and companies will begin to spend more and more of their dollars and euros on buying up properties outside Russia. While a few might buy up sports teams, as Roman Abramovich did when he bought the Chelsea soccer team in London, others will expand their holdings of manufacturing and service companies as LUKoil did when it bought up the Getty Oil filling stations network and Nelson Resources, which had major oil holdings in Kazakhstan. Generally Russians purchase foreign companies that produce products similar to what they produce within Russia.[58] On August 8, 2007, the *Financial Times* reported that Oleg Deripaska, one of the oligarchs who has worked closely in the past with Putin, had invested $1.54 billion to purchase a large stake in Magna, a Canadian auto parts manufacturer. Even more spectacular, Deripaska had bought up almost 5 percent of General Motors stock for $900 millon. The assumption is that he planned to build on this link with General Motors to upgrade the technology and work practices at Gaz, his Russian automobile manufacturing plant.

There will be people who oppose such investments, especially in companies that have strategic significance such as Stillwater Mining, the United States' only producer of platinum and palladium. The Germans under the leadership of Angela Merkel seem to be particularly sensitive to such "sovereign investment fund" initiatives. They fear that not only the Russians but also the Chinese will use their huge foreign currency reserves to acquire equity in defense-related companies as well as recently privatized industries. According to the spokesman for the German government, while "private investors would generally be welcome," the Germans were committed to preventing state sovereign investment funds as well as Russian and Chinese nationalized industries from buying up German businesses. In the words of Roland Koch, the German official sponsoring such legislation, "We didn't just go through all our efforts to privatize industries like Deutsche Telekom or the Deutsche Post only so that the Russians can nationalize them." Along the same lines, the European Commission, the executive office of the European Union, has proposed legislation that will require energy companies to unbundle or separate their energy-producing divisions from the units that distribute and transport that energy. This, the European Commission argues, would stimulate competition and prevent a single gas supplier to a region from dominating the distribution network. Such legislation will make it difficult for a company like Gazprom to institute an embargo and generally monopolize control. Subsequently there have also been proposals that would require members of the EU to notify the EU before they conclude any bilateral agreement with third parties (read Russia) that would affect EU interests.[59]

The European Commission of the European Union on September 19, 2007, adopted a resolution insisting that no company from outside the EU be allowed to acquire energy infrastructure (i.e., gas distribution companies) "in Europe unless there is 'reciprocity with that country.'"[60] It was assumed this was also intended specifically to prevent further acquisitions within the EU by Gazprom; certainly some members of the Russian Duma interpreted it that way and were not happy about it.

Others take an opposite approach and may actually welcome Russian efforts to buy up local natural gas distribution service companies and properties, much as it has done in Belarus, Lithuania, and Germany. As they see it, if the Russians own these distribution facilities, they will be less likely to shut off the flow of gas to their own facilities.[61] On the negative side, control of such assets will allow Russian interests to exclude non-Russian-originated gas from their pipeline. They may even be able to extract monopoly rents. At the same time, foreign assets owned by Russians outside of Russia risk ending up as hostages. Presumably the rule of law that prevails in the United States and other Western countries will prevent retaliation and the arbitrary seizing of property owned by Russians. As we saw earlier, however, Noga, the Swiss company, managed to win permission from a European judge to seize the Luxembourg bank owned by the Russian government, suggesting that the more property owned by Russian entities outside Russia, especially where the state is the dominant owner, the more vulnerable Russia becomes to the type of pressure and blackmail that on occasion Western companies in Russia have faced.

This implies that U.S. policy should encourage Russian companies to invest in the United States, especially when they provide goods and services that supplement those sold by others. In the same way, the more assets owned by Russian entities outside of Russia, the more Russian firms are likely to feel pressure to conform to international standards. Thus, if Gazprom should for some reason decide to withhold delivery of LNG to its U.S. customers, as a countermeasure those customers might well decide to seize Russian-owned assets as a hostage. In the same way, companies like LUKoil should be encouraged to compete in U.S. markets by exporting Russian petroleum to the gasoline service stations that it owns in the U.S. (This is also a way to diversify oil imports.) When Russia is able to act as a monopolist, however, such investments are more problematic—building a natural gas pipeline that gives them monopoly power over consumers in the territory served by that pipeline would be an example.

THE NEW ASSERTIVE RUSSIA

In the aftermath of the financial collapse of August 1998, it looked as if Russia's day as a superpower had come and gone. That it should recover and reassert itself again after less than a decade is nothing short of an economic and political miracle.

Yet its recovery is heavily dependent on high energy prices, making that recovery precarious. Countries with a monoculture based on energy more often than not have complicated social, political, and economic problems caused in part by their overreliance on energy resources. Nevertheless, most Russians would agree that the benefits that stem from that monoculture have more than offset the disadvantages. Russian petroleum export revenue has provided Russia with extraordinary liquidity as a creditor nation—a rather unusual, if not historic, status for Russia. Moreover, its natural gas wealth and its ownership of natural gas pipelines, at a time when energy markets are likely to be severely constrained for years to come, give Russia unprecedented political and economic power. Whether Russia joins in a strong GECF (Gas Exporting Countries Forum) consortium is irrelevant; because of its pipeline operations, countries served by this pipeline, be they in Europe or Asia (including China), will have a hard time resisting Russian demands. This is very different from the MAD (mutually assured destruction) days of the Cold War when each side was afraid of the other. For the foreseeable future, there seems to be no way to restrain Russian behavior. Yet as they pile up more and more dollars and euros, the Russians are almost certain to use more and more of their money to expand and buy up assets overseas. Once they are exposed this way, they in turn may find themselves held hostage.

For now, however, Russia, with or without Vladimir Putin as president, finds itself in a newly assertive, even dominant, international position. Its emergence as a new super energy power overlaps with the weakening of the United States as the country has squandered its manpower and resources in Iraq. Russia under Putin, on the other hand, has developed a new hubris that is not based on mere bluster.

This has manifested itself in several ways. For instance, with its newfound cash, Russia once again is increasing the funding of its military. In 2005, military expenditures rose 27 percent over 2004 and in 2006, they were increased by another 22 percent. And in an episode reminiscent of the nineteenth century, Russia sent an expedition in two mini-submarines 15,000 feet under the Arctic ice cap. The submarines planted a titanium capsule with a Russian flag under the North Pole and then claimed ownership of 460,000 square miles of Arctic Ocean floor. Asserting that the 1,240-mile underwater Lomonosov Mountain Range extends from the Russian mainland into the polar region, Russian geologists took core samples of the ground at the Pole which they insisted matched similar samples from the underwater area part of Russia. Their reasoning is that this makes the North Pole part of Russia. This not only has military implications but the region is thought to contain as much as 10 billion tons of oil and gas deposits, equal to perhaps 25 percent of the world's as yet undiscovered oil and gas; and with earth warming, these energy deposits are now much more readily accessible.[62]

A land claim of this sort by the Russian government is not all that novel. Czarist governments claimed parts of China as well as Alaska in the nineteenth century and the Soviet government regularly presented claims over territories of other countries including Finland, Rumania, Poland, and Japan. What is unprecedented now is that with its new financial holdings, Russia has also begun to demand a say in world monetary and financial decision making. In a June 2007 speech to a group of Western and Russian businessmen, Putin went so far as to challenge the way the World Trade Organization and the International Monetary Fund (IMF) operate. He called them "archaic, undemocratic and awkward," complaining that the United States, the EU, and Japan run the organization as if it were a privileged private club. Putin wanted more say for Russia and other non-Western countries. With that in mind, a few months later Russia proposed that Josef Tosovsky, a well-regarded Czech central banker and former prime minister, be made the head of the IMF instead of the EU's nominee, Dominique Strauss-Kahn, the former French finance minister. What made this so newsworthy is that by tradition, Europeans and now the EU select the head of the IMF while the United States nominates the head of the World Bank. Moreover, in the Soviet era, the USSR more or less ignored the World Bank and the IMF. Now that modern-day Russia is a financial power-house, Putin has decided to challenge such cozy arrangements. In the same St. Petersburg speech, Putin questioned why only the dollar and the euro serve as world currencies. Given the new strength of the ruble, Putin argued that the ruble should serve the same function.

These are clearly good times for Russia. In many ways it is like awakening from a decade-long nightmare. Perhaps we in the West did not fully appreciate how upset most Russians were after the loss of their country's superpower status. It was a national disgrace and humil-iation. How deeply this was felt can be gauged by revealing comments from two Russian political figures. Reacting to the British govern-ment's demand that the Russian government turn over Andrei Lugovoi, a former KGB agent, to them so they can question him about the polo-nium 210 poisoning of Alexander Litvinenko, Konstantin Kosachëv, chairman of the Russian Duma's Foreign Affairs Committee, exploded. Normally a reasonable fellow, Kosachëv protested, "You can act this way toward a banana republic, but Russia is not a banana republic." In an even more revealing statement, one better suited for a psychiatrist's couch than the weekly newspaper, *Argumenty i fakty*, in which it appeared, Valentina Matviyenko, the governor of St. Petersburg, proudly insisted that "Russia has now regained a sense of self-respect. We spent so many years feeling there was something wrong with us—others lecturing us on how we should live and where we should go. But we have overcome our inferiority complex."

Putin clearly shares this new assertiveness. This was dramatically demonstrated in Munich, Germany, on February 10, 2007, at the Annual Munich Conference on Security Policy. In a speech to the group of mostly NATO member countries, Putin criticized not only NATO for opening bases on what used to be Warsaw Pact territory—something NATO promised it would not do—but also President George W. Bush and the United States for pursuing unilateral policies in Iraq and even Eastern Europe. The tone and the content of his speech to representatives of countries that were once enemies of the Soviet Union would have been unthinkable nine years earlier. It was a tough speech that unnerved many Western representatives who had seldom, if ever, heard such candor in a public attack on the United States at a meeting of America's closest allies.

Putin's speech reflected Russia's progress in the short time since Russia had become a major energy producer. It was not the tough part of the talk, however, but the softer part that best demonstrated Russia's new assertiveness and self-confidence. After having taken President Bush to task in such an open way, during the question period Putin suddenly softened his mood. "In spite of all our disagreements, I consider the president of the United States my friend. . . . He is a decent person, and it is possible to talk and reach an agreement with him"; or as the *New York Times* translated Putin's Russian, Bush "is a decent man and one can do business with him."[63]

For those with long memories, this condescending sentiment all but echoed what then Prime Minister Margaret Thatcher said in December 1984 after she first met that "unwashed" Politburo yokel from Moscow, Mikhail Gorbachev. This was but three months before he would become the leader of the USSR. Now twenty years later as the head of this new energy powerhouse, it was Putin's turn to show a similar condescending manner. Whether he meant it or not—and he probably meant it—Putin now wanted to show that the tables have turned. It was Russia's turn to be condescending to that "unwashed" Texas yokel, George W. Bush. There was not much the targets of his attack could do but smile and seek to import more of Russia's gas and oil.

Notes

Introduction

1. *SKEIN News*, November 30, 2005.
2. *Financial Times*, April 24, 2006, 2.
3. *David Johnson's Russia List*, no. 26 (March 3, 2007).
4. *Eurasia Daily Monitor*, Jamestown Foundation, September 21, 2006; *Moscow Times*, March 23, 2007.
5. John D. Grace, *Russian Oil Supply: Performance and Prospects* (Oxford: Oxford University Press, 2005), 16.
6. Marshall I. Goldman, "The Russian Disease," *International Economy* (Summer 2005), 27.

Chapter 1

1. Robert W. Tolf, *The Russian Rockefellers* (Stanford: Hoover Institution Press, 1976), 40.
2. G. Segal, "The Oil and Petrochemical Industry in the Soviet Union" (London: mimeo., undated), 1.
3. Tolf, 41–42.
4. Iain F. Elliot, *The Soviet Energy Balance* (New York: Praeger, 1974), 70.
5. Elliot, 70.
6. Daniel Yergin, *The Prize* (New York: Simon & Schuster, 1991), 57.
7. Segal, 2; V. I. Pokrovskii, *Sbornik Svedenii Po Istorii I Statistike Vneshnei Torgovli Rossii* (St. Petersburg: Department tamozhennykh sbor', 1902), Vol. I, 217.
8. Pokrovskii, 217.
9. Elliot, 70; Tolf, 44.
10. Tolf, 40. This is an excellent study of Nobel's life and work. Much of the description of Nobel's activities in the remainder of this chapter is taken from this book and the study by Pokrovskii. See also Steve Le Vine, *The Oil and the Glory* (New York: Random House, 2007), chapters 1–3.

11. Pokrovskii, 218.
12. Tolf, 55.
13. Pokrovskii, 219.
14. Segal, 3.
15. Pokrovskii, 220.
16. Tolf, 97.
17. Tolf, 95.
18. Pokrovskii, 221.
19. Tolf, 48, 71–72, 99.
20. Pokrovskii, 221.
21. U.S. Bureau of the Census, Historical Statistics of the United States, Colonial Times to 1970, bicentennial ed., Pt. I (Washington, D.C., 1975), 597.
22. Pokrovskii, 215.
23. Pokrovskii, 142.
24. Pokrovskii, 215; Peter Lyashchenko, History of the National Economy of Russia to the 1917 Revolution (New York: Macmillan, 1949), 689.
25. Zbynek Zeman and Jan Zoubek, Comecon Oil and Gas (London: Financial Times, 1977), 9; Tolf, 188–92.
26. Tolf, 192–93.
27. Pokrovskii, 217; Segal, 2.
28. Yergin, 131.
29. Yergin, 133.
30. Lyashchenko, 632.
31. Exports in 1913, however, did slightly exceed those in 1905 by 3,000 tons.
32. Trudy statisticheskogo otdeleniie, Departament tamozhenny sbor', Obzor' Vneshnei torgovli Rossii za 1901 (St. Petersburg: Department tamozhenny sbor', 1903), 9 (hereafter Vneshnei torgovli Rossii and the appropriate year); Vneshnei torgovli Rossii 1915, 8.
33. A. V. Venediktov, Organizatsiia gosudarstvennoi promyshlennosti v SSSR (Leningrad: Izdatel'stvo Leningradskogo Universiteta, 1957), 185.
34. Tolf, 215.
35. Tolf, 217.
36. Segal, 5.
37. Anthony C. Sutton, Western Technology and Soviet Economic Development, Vol. I, 1917 to 1930 (Stanford: Hoover Institution Press, 1968), 41; Tolf, 222.
38. Sutton, Vol. I, 18.
39. Sutton, Vol. I, 19, 30, 122; the Japanese did not come in until 1925.
40. Sutton, Vol. I, 37.
41. Sutton, Vol. I, 10, 23.
42. Sutton, Vol. I, 82.
43. Venediktov, Vol. I, 185.
44. Segal, Vol. I, 16.
45. Foreign Trade, September 1967, 18.
46. Sutton, Vol. I, 41.

47. *Foreign Trade*, September 1967, 17.
48. Some of that might not have gone to the United States directly but to Standard Oil for reshipment to the Middle and Far East.
49. Ministersvo vneshnei torgovli SSSR (hereafter MVT SSSR), *Vneshniaia torgovlia SSSR za 1918–1940* (Moscow: Vneshtorgizdat, 1960), 657 (hereafter *VT SSSR* with the appropriate year).
50. *VT SSSR, 1918–1940*, 528.
51. Sutton, Vol. I, 16, 40. Later on Sutton refers specifically to 1928 when he says petroleum amounted to 19.1 percent of all earnings. His assertions are then repeated by Tolf and Segal.
52. *VT SSSR, 1914–1940*, 94.
53. Segal, 12.
54. Elliot, 72; Grace, 14.
55. Elliot, 74–75, 89; Robert Campbell, *The Economics of Soviet Oil and Gas* (Baltimore: Johns Hopkins University Press, 1968), 124–25.

Chapter 2

1. Pravda, December 10, 1963, 1; *VT SSSR, 1913–1940*, 58, 144.
2. See Table 1.2; Elliot, 74; Anthony C. Sutton, *Western Technology and Soviet Development*, Vol. II, 1930 to 1945 (Stanford: Hoover Institution Press, 1971), 90.
3. Sutton, Vol. II, 89; Anthony C. Sutton, *Western Technology and Soviet Development*, Vol. III, 1945 to 1965 (Stanford: Hoover Institution Press, 1973), 38–39, 135–36.
4. Campbell, *Economics*, 126.
5. Campbell, *Economics*, 127–28.
6. Campbell, *Economics*, 125; Robert W. Campbell, *Trends in the Soviet Oil and Gas Industry* (Baltimore: Johns Hopkins University Press, 1976), 27.
7. Elliot, 96; "Europe-Sibir," *Ekonomika i organizatsiia promyshlennogo proizvodstva* (hereafter *EKO*), no. 4, 1976, 160–67.
8. *Total Information*, no. 68, 1976, Geneva, Switzerland, 2.
9. *Total Information*, 4; Elliot, 97; *The Review of Sino-Soviet Oil*, April 1979, 50.
10. Sutton, Vol. I, 24; Tolf, 65, 182.
11. Campbell, *Economics*, 102–20, 126–28; Joseph S. Berliner, *The Innovation Decision in Soviet Industry* (Cambridge: MIT Press, 1976).
12. Campbell, *Economics*, 93.
13. Campbell, *Economics*, 109.
14. Campbell, *Economics*, 93.
15. *Pravda*, February 28, 1978, 2.
16. *Review of Sino-Soviet Oil*, November 1977, 10.
17. *Current Digest of the Soviet Press*, February 14, 1973, 6.
18. *Pravda*, November 20, 1962, 4.
19. *Izvestia*, October 15, 1984, 47.
20. *Pravda*, January 27, 1978, 2; Daniel Park, *Oil and Gas in COMECON Countries* (London: Kogan Page, 1979), 67.

21. *Turkmenskaia iskra*, December 6, 1977, 2; *Literaturnaia gazeta*, January 18, 1978, 10.

22. *CIA Gas*, July 1978, 2.

23. Marshall I. Goldman, *USSR in Crisis* (New York: W. W. Norton, 1983), 66.

24. Marshall I. Goldman, *The Enigma of Soviet Petroleum, Half Full or Half Empty* (London: George Allen & Unwin, 1980).

25. Goldman, *USSR in Crisis*, 136–38.

26. Goldman, *USSR in Crisis*, 144.

27. Yergin, 503.

28. *MIG Soviet Foreign Aid* (New York: Praeger, 1967), 97.

29. *Moscow Narodny Bank Bulletin*, February 4, 1971, 2.

30. Yergin, 50; Richard Matzke, "Russia and the USA: No Longer Rivals, Not Yet Partners," *Demokratizatsiya*, 15, no. 4 (Fall 2007), 371.

31. *Financial Times*, November 27, 2006, 4.

32. Peter Schweizer, *Victory* (New York: Atlantic Monthly Press, 1994).

33. Schweizer, 219.

34. Schweizer, 242.

35. Yegor Gaidar, *Gibel' Imperii* (Moscow: Rosspen, 2006), 224–31; Yegor Gaidar, "The Collapse of the Soviet Union: Lessons for Contemporary Russia," seminar, American Enterprise Institute, Washington, D.C., November 13, 2006.

36. Gaidar, *Gibel*, 227.

37. Gaidar, *Gibel*, 286, 287, 328.

38. Graham Allison and Robert Blackwell, "America's Stake in the Soviet Future," *Foreign Affairs*, 70, no. 3 (Summer, 1991), 77–97.

39. Gaidar, *Gibel*, 270, 315, 337.

40. Gaidar, *Gibel*, 337.

41. Gaidar, *Gibel*, 244.

42. Schweizer, xii.

43. Gaidar, *Gibel*, 223, 310.

44. Gaidar, *Gibel*, 204, 213, 337.

Chapter 3

1. "The Unofficial Economy in Transition," *Brookings Papers on Economic Activity*, no. 2, 1997, 159.

2. *The Moscow Times*, May 8, 2001.

3. Marshall Goldman, *The Piratization of Russia: Russian Reform Goes Awry* (New York: Routledge, 2003), 106.

4. Goldman, *The Piratization of Russia*, 117.

5. Nina Poussenkova, "Rosneft as a Mirror of Russia's Evolution," *Pro et Contra Journal*, 10, no. 2 (June 2006), 11.

6. *Washington Post*, August 22, 2001; Chrystia Freeland, *Sale of the Century* (Toronto: Doubleday, 2000), 175.

7. Goldman, *The Piratization of Russia*, 150; *Moscow Times*, April 4, 2000; *Moscow News*, June 28, 1999, 2.

8. *Moscow Times*, April 11, 2000; Goldman, *The Piratization of Russia*, 151.
9. *Moscow Times*, April 25, 2006.
10. Goldman, *The Piratization of Russia*, p.143.
11. Interview, December 8, 2000.
12. Goldman, *The Piratization of Russia*, 126.
13. Goldman, *The Piratization of Russia*, 119.
14. *New York Times*, November 21, 2006, C11.
15. *Observer*, January 8, 2006, Business section,1.
16. U.S. Court of Appeals for the Second Circuit, August Term, 2004, Docket No. 04–1357-cv, 8–9.
17. Goldman, *The Piratization of Russia*, 144.
18. *Wall Street Journal*, March 24, 2007, A3; *Business Week*, April 19, 2007; *Financial Times*, March 24, 2007, 9.
19. *Wall Street Journal*, September 15, 2006; *Russian Analytical Digest*, August 2006, 11.
20. *New York Times*, November, 24, 2002, F1.

Chapter 4

1. Marshall Goldman, "Russia's Bleeding Heartland," *Central European Economic Review*, September 1997, 6.
2. Anders Aslund, *How Russia Became a Market Economy* (Washington: Brookings, 1995); Richard Layard and John Parker, *The Coming Russian Boom* (New York: Free Press, 1996).
3. *Boston Globe*, October 28, 1998, A8.
4. *Financial Times*, July 17, 2003, 14.
5. *Wall Street Journal*, October 23, 1998, A3.
6. Grace, 80, 83.
7. *Wall Street Journal*, May 24, 2006, A6.
8. Zhou Dadi, "Sustainable Energy Development in China: Challenge and Opportunities," March 21, 2007, mimeograph, 1.
9. *Financial Times*, May 2, 2007, 10.
10. Dadi, 1.
11. *Financial Times*, July 4, 2006, 3
12. Dadi, 2: ITAR TASS, December 6, 2006.
13. *China Daily*, September 18, 2007, 1; *Financial Times*, November 30, 2007, 4.
14. *Financial Times*, July 19, 2007, 7.
15. *Wall Street Journal*, October 7, 2006.
16. Grace, 166; Mikhail Khodorkovsky, Presentation to German Electricity Association Congress, 2000, at www.yukos.com/exclusive/exclusive.asp?ID=6085.
17. *David Johnson's Russia List*, January 2, 2007, No. 1, Items 6, 7.
18. *Moscow Times*, September 7, 2006, 7; Prime-TASS, September 1, 2006.
19. *Moscow Times*, August 26, 2006; *Financial Times*, June 17, 2006, 1.
20. *Rodnaya Gazeta*, no. 38, October 5, 2006; *Moscow Times*, September 7, 2006; August 25, 2006.

21. "The Russian Economy," World Bank, April, 2006, Moscow Office, 9.

22. *Financial Times*, July 26, 2006, 21.

23. *Wall Street Journal*, January 22, 2007, A1.

24. *Gazprom in Questions and Answers* (Moscow: Gazprom, 2006), 6.

25. *David Johnson's Russia List*, January 2, 2003, 11.

26. *Moscow Times*, December 5, 2000; *New York Times*, November 24, 2002, 4; *Moscow News*, June 8, 2005, 9.

27. *International Herald Tribune*, July 24, 2004.

28. *New York Times*, November 24, 2002, F1; *Moscow Times*, September 24, 1997; *Financial Times*, September 1, 1994, 2.

29. *Russian Journal*, November 29, 1999.

30. *Wall Street Journal*, September 1, 1994, A6.

31. Annual Arden House Seminar, Liz Williamson, Conoco, March 18, 1995.

32. *Moscow Times*, December 5, 2000; *Houston Chronicle*, December 30, 2004.

33. Discussion with senior ConocoPhillips executives, September 10, 2003.

34. Initially, in September 2004 it bid $2 billion to buy the 7.5 percent of LUKoil's share held by the government. By March 2005 it had acquired 11.3 percent; three months later, 12.6 percent; and by December 3, 2005, it owned 16.1 percent. It purchased the remaining 3.9 percent in 2006 for $3 billion, for combined expenditures of over $7.5 billion. *New York Times*, June 3, 2005, C4; Prime-TASS, October 3, 2006.

35. *Moscow Times*, August 30, 2004, 5.

36. *New York Times*, October 28, 2004, C8.

37. *David Johnson's Russia List*, June 5, 2007, no. 127, 135.

38. *New York Times*, December 1, 2005, C1, 8.

39. *Moscow Times*, October 4, 2006; *Vedomosti*, October 24, 2005; *David Johnson's Russia List*, October 25, 2005, 9277, item #9.

40. *FC Novosti*, Ia, October 10, 2007, 12:22; *Financial Times*, October 11, 2007, 2.

41. *Wall Street Journal*, January 30, 2007, A3.

42. *Asian Wall Street Journal*, November 8, 2006, 4.

43. *Insight TNK-BP International News Letter*, Autumn 2006, 2.

44. *Financial Times*, July 4, 2006, 3.

45. *Financial Times*, September 12, 2006, 15; October 12, 2006, 8.

46. *Wall Street Journal*, January 25, 2007, A12; International Energy Agency Outlook, 2006.

47. *Financial Times*, October 16, 2006, 13.

48. Bernard A. Gelb, "Russian Oil and Gas Challenges," *CRS Report for Congress*, Library of Congress, Washington, January 3, 2006, CRS 2; Taken from *BP Statistical Review of World Energy*, 2005 and 2006, 6; *Oil and Gas Journal*, December 2004. John Grace estimates the Russians have proven reserves of 68.2 billion barrels. Grace, 179.

49. *Wall Street Journal*, September 30, 2004, A1; *New York Times*, October 28, 2004, C8.

50. *New York Times*, October 28, 2004, C8.

51. The careful geologist John Grace tends to be considerably more skeptical. Grace, 216.

52. *Prime-TASS Business Newswire*, April 26, 2007; *Financial Times*, April 25, 2007, 176 *Moscow Times*, August 13, 2007.

53. Jonathan P. Stern, *The Future of Russian Gas and Gazprom* (Oxford: Oxford University Press, 2005), 144.

Chapter 5

1. For a more thorough discussion as well as a translation of an article Putin wrote on this theme, see Harley Balzer, "Vladimir Putin's Academic Writings and Russian Natural Resource Policy," *Problems of Post-Communism* (January/ February 2006), 48–54. See also Robert Price, "Putin's National Champions: Domestic and International Implications," Master's Thesis, Harvard University, August 2007, 6–8.

2. Balzer, 51.

3. Balzer, 52.

4. Balzer, 53.

5. Balzer, 54.

6. *Washington Times*, March 31, 2006; *Pittsburgh Tribune Review*, March 28, 2006.

7. *Kommersant*, April 6, 2006.

8. *Financial Times*, May 4, 2007, 18.

9. Stern, 170.

10. Stern, 170.

11. Stern, 170.

12. Interview with Boris Berezovsky, October 22, 2003, London.

13. Grace, 171.

14. Grace, 120; *Moscow News*, March 24, 2004, 9.

15. *Financial Times*, December 20, 1999, 11. Not by any means unique to Yukos, transfer pricing was a widely used practice among Russian oil companies, including Rosneft, the state-owned company that was to be the main beneficiary of Yukos's collapse. *Kommersant*, February 6, 2007, 1.

16. Prime-TASS, March 6, 2006.

17. *Moscow Times*, April 25, 2006; July 24, 2004; August 7, 2007; *BBC*, July 26, 1998; *Financial Times*, July 27, 2004, 1, 9; *Times of London*, August 9, 2004; *Independent*, July 27, 2004, 22.

18. *Moscow Times*, April 25, 2006.

19. *Moscow Times*, April 25, 2006.

20. *Wall Street Journal*, August 26, 1999, A1; September 3, 1999, 2; *Moskovskaia Pravda*, December 17, 1994, 5.

21. *Moscow Times*, January 21, 2002.

22. Grace, 121.

23. Seminar, John Pappalardo, lawyer for Khodorkovsky, October 25, 2005; *Wall Street Journal*, April 27, 2004, A1.

24. *Moscow Times*, October 25, 2006.
25. *Wall Street Journal*, April 27, 2004, A12.
26. *Wall Street Journal*, April 27, 2004, A12.
27. *Wall Street Journal*, April 12, 2004, A12.
28. *New York Times*, February 11, 2007, A4.
29. Nina Poussenkova, *Pro et Contra Journal*, 10, no. 2 (June 2006).
30. Bruce Bean, "Russia's Yukos Affair: The Use and Abuse of Law," photocopied 2005, 5.
31. *Financial Times*, November 9, 2005, 16.
32. *Financial Times*, November 9, 2005, 16; *Vedomosti*, February 20, 2003.
33. Interview at Davis Center, Harvard University, December 6, 2004.
34. Kompromat.ru, July 4, 2003.
35. *Moscow News*, March 3, 2004, 3; March 24, 2004, 9.
36. *Moscow Times*, July 15, 2003.
37. Seminar, Davis Center, October 25, 2005.
38. *Moscow Times*, February 27, 2007.
39. *Kommersant*, April 25, 2006.
40. Mikhail Khordorkovsky, "What Is the Morality Tale?" Panel Discussion at the University of Pennsylvania Law School, December 7, 2006.
41. Seminar, Davis Center, Harvard University, October 25, 2005; Panel Discussion, University of Pennsylvania Law School, December 7, 2006.
42. *Moscow Times*, February 27, 2007.
43. *Financial Times*, March 26, 2007, 16; *Kommersant*, February 6, 2007, 1.
44. *New York Times*, March 17, 2007, A3.
45. *Moscow Times*, April 5, 2007.
46. Prime-TASS, February 15, 2007.
47. *Moscow Times*, February 15, 2007; *Wall Street Journal*, February 22, 2007, A4.
48. *Wall Street Journal*, February 22, 2007, A4.
49. *Financial Times*, March 26, 2007, 16; August 20, 2007; *Moscow Times*, July 23, 2007, A4.
50. *Moscow Times*, March 28, 2007; *New York Times*, March 28, 2007, C3.
51. *New York Times*, March 27, 2007, A3; March 28, 2007, C3; *Wall Street Journal*, March 28, 2007, A10.
52. *Moscow Times*, March 28, 2007.
53. *Moscow Times*, March 28, 2007.
54. *Wall Street Journal* July 12, 2007, A7; *Financial Times*, January 29, 2008, 18.
55. *Moscow Times*, February 27, 2007.
56. *Financial Times*, March 12, 2004.
57. *Financial Times*, October 11, 2005, 5; *Washington Post*, March 4, 2004.
58. *Financial Times*, October 11, 2005, 5.
59. *Independent*, June 18, 2004, 1.
60. *Wall Street Journal*, July 31, 2007, A2.
61. *Moscow Times*, August 1, 2007.
62. *Moscow Times*, March 7, 2006.

63. *Financial Times*, November 4, 1998, 2; *Moscow Times*, April 30, 2004; July 21, 2006; Grace, 136.
64. *New York Times*, July 3, 2004, B1.
65. Phone conversation with Jeff Larson of Sowood Capital, agent for the Harvard Management Company, February 28, 2007.
66. *Business Week*, October 23, 2006, 52.
67. Grace, 142.
68. *Financial Times*, March 22, 2007, 18.
69. *Wall Street Journal*, March 24, 2007, A3; *Financial Times*, March 24, 2007, 1.
70. *Moscow Times*, October 16, 2006; February 27, 2007. They have also been accused of pollution and the alleged cutting of trees while building pipelines.
71. *Financial Times*, February 28, 2007, 21; *Moscow Times*, March 1, 2007.
72. *New York Times*, October 6, 2006, C4.
73. *Wall Street Journal*, September 22, 2006, A6; *Financial Times*, September 22, 2006, 2.
74. *Financial Times*, April 20, 2007, 4: Moscow Times, June 9, 2007, 6.
75. *Financial Times*, September 22, 2006, 2; Fiona Hill and Florence Fee, "Fueling the Future: The Prospects for Russian Oil and Gas," *Demokratizatsiya* (Fall 2002), 481.
76. Prime-TASS, March 1, 2007; *Moscow Times*, June 22, 2007.
77. *Moscow Times*, March 1, 2007; February 27, 2002.
78. *Financial Times*, February 28, 2007, 21; March 22, 2007, 18.
79. *Forbes.com*, August 23, 2007, 11:50 a.m.
80. Grace, 194.
81. *Moscow Times*, February 27, 2007.
82. *Moscow Times*, March 23, 2007.
83. *Financial Times*, September 22, 2004, 21; August 5, 2005, 14.
84. Yergin, 456, 583.

Chapter 6

1. Yergin, 742–43.
2. Yergin, 742–43; Stern, 215.
3. Yergin, 743.
4. Yergin, 743.
5. Balzer, 48.
6. Bernard A. Gelb, "Russian Oil and Gas Challenges," Congressional Research Service Report for Congress, Library of Congress, January 3, 2006; *Sunday Times*, July 9, 2006, 24; *BP Statistical Review of World Energy*, 2006, 22; *Gazprom in Questions and Answers*, 20.
7. *BP Statistical Review*.
8. *Financial Times*, May 16, 1997, 3; May 22, 1997, 18; May 27, 1997, 1; *Russia Review*, July 2, 1997, 23.
9. See Introduction.
10. *Moscow Times*, May 29, 2001.
11. Interview, June 21, 2001.
12. *Moscow Times*, March 13, 2007; *New York Times*, July 13, 2006, 10; Stern, 106.

13. *Moscow Times*, March 13, 2007.
14. *Moscow Times*, May 26, 2006.
15. President Putin's State of the Nation Speech to Duma, May 10, 2006.
16. Valdai Hills Discussion Group, September 4, 2006.
17. Gazpromistan was coined by Edward Lucas, a writer for *The Economist*.
18. Prime-TASS, March 1, 2007.
19. *Financial Times*, July 14, 2006, 9.
20. *Financial Times*, April 28, 2006, 2.
21. *Financial Times*, July 14, 2006, 9.
22. *Novaia Gazeta*, August 21, 2006; *Kompromat.ru*, "*Samikh Bogatikh na Gaze*"; *Financial Times*, April 27, 2006, 2; April 28, 2006, 2.
23. *Financial Times*, April 28, 2006, 2.
24. *Wall Street Journal*, April 21, 2006, A6; September 22, 2006, 4.
25. *Wall Street Journal*, September 22, 2006, A4.
26. *Wall Street Journal*, September 22, 2006, A4.
27. *Moscow Times*, April 24, 2007.
28. *Moscow Times*, October 4, 2006.
29. *Moscow Times*, October 4, 2006.
30. *Wall Street Journal*, December 23, 2006, A4; *Financial Times*, December 23, 2006, 2.
31. *Moscow Times*, March 1, 2007. Seminar, Davis Center, December 8, 2006, Zurab Noghaideli, Prime Minister of Georgia.
32. *Moscow Times*, March 20, 2007; *Eurasia Daily Monitor*, Jamestown Foundation 4, no. 97 (May 17, 2007), 2.
33. *Financial Times*, February 28, 2007, 8.
34. *Financial Times*, December 14, 2006, 1.
35. *BBC Monitoring*, January 4, 2007, 13:00–1.
36. *New York Times*, January 13, 2007, A4.
37. *Financial Times*, August 3, 2007, 1.
38. *Financial Times*, December 14, 2006, 1; *Rossiiskaia gazeta*, May 21, 2007, 1.
39. *Eurasia Daily Monitor*, Jamestown Foundation 3, no. 57 (March 23, 2006), 4; *Eurasia Daily Monitor*, Jamestown Foundation 3, no. 190 (October 16, 2006), 1.
40. *Financial Times*, May 4, 2006.
41. *New Europe*, August 2, 2006.
42. *New Europe*, April 25, 2007.
43. *Financial Times*, December 14, 2006, 3.
44. *Wall Street Journal*, December 4, 2006, A6.
45. *Eurasia Daily Monitor*, Jamestown Foundation 4, no. 140 (July 19, 2007).
46. *Eurasia Daily Monitor*, Jamestown Foundation 4, no. 51 (March 14, 2007).
47. *Eurasia Daily Monitor*, Jamestown Foundation 3, no. 174 (September 21, 2006); 4, no. 51 (March 14, 2007).
48. *Agence France-Presse*, March 14, 2007.
49. *Moscow Times*, March 23, 2007.
50. *Vedomosti*, June 25, 2007, B3.

51. *Eurasia Daily Monitor,* Jamestown Foundation 4, no. 144 (July 25, 2007), 4; *Financial Times,* August 20, 2007, 14.
52. *Moscow Times,* February 15, 2007. To broaden the base, both German companies also agreed to transfer 4.5 percent of their holdings to Dutch Gasunie.
53. *Russia Profile* 4, no. 2 (March 2007), 48.
54. *Wall Street Journal,* March 31, 2006, A15.
55. *Financial Times,* April 3, 2006, 2.
56. *Moscow Times,* April 3, 2006.
57. *Financial Times,* April 3, 2006, 2.
58. *Financial Times,* April 15, 2006.
59. Stern, 144.
60. *Eurasia Daily Monitor,* Jamestown Foundation 4, no. 160 (August 16, 2007), 2.
61. *Moscow Times,* March 31, 2006.
62. *Financial Times,* April 22, 2005, 14.
63. *Mosnews.com,* February 24, 2005; *Financial Times,* December 22, 2005, 14.
64. *New Europe,* February 28, 2007.
65. *Business Wire,* Prime-TASS, April 10, 2007; *Eurasia Daily Monitor,* Jamestown Foundation 4 (April 23, 2007); *Financial Times,* September 21, 2007, 3.
66. *Financial Times,* December 12, 2007, 6.
67. *The Economist,* February 22, 2007.
68. Kennan Institute Meeting Report 24, no. 3 (2006).
69. http://HRRP;/SVT.SE/SVT/JSP/CROSSLINK. JSP?D=53332&A=717462.119=SENASTENYTT_613854&1POS=RUBRIK_717462E
70. *Eurasia Daily Monitor,* Jamestown Foundation 4, no. 68 (April 6, 2007).
71. *Eurasia Daily Monitor,* Jamestown Foundation 4, no. 53 (March 16, 2007); *New Europe,* January 24, 2007.
72. *Eurasia Daily Monitor,* Jamestown Foundation 4, no. 53 (March 16, 2007).
73. A study prepared jointly by the Council on Foreign and Defense Policy and the State University, Higher School of Economics in Moscow argued that the geo-political-energy situation "in the Caspian region is generally developing in favor of the West." Meeting with criticism, the authors decided to revise their findings to show a greater likelihood of Russian progress and they dropped the conclusion that "Russia's influence in the Caspian region will be minimized." The Council on Foreign and Defense Policy, State University, Higher School of Economics, RIO-Center, *The World around Russia: 2017: An Outlook for the Midterm Future,* Moscow, 2007, 236; *Vremya Novosti,* January 11, 2007, 2; *Current Digest of the Post-Soviet Press,* January 31, 2007, 5.
74. *Asia Times Online,* May 27, 2006.
75. *Financial Times,* January 23, 2007, 19; August 20, 2007, 6.
76. Stern, 144.
77. This is not unique to Russia. The U.S. government among others has used similar tactics as when it imposed an economic blockade against Cuba and earlier against China.
78. *Financial Times,* May 30, 2006, 4.

79. *Financial Times*, November 27, 2006.

80. *Rossiyskaia Gazeta*, November 29, 2006.

81. Kremlin, Russia, February 1, 2007.

82. Stern, 144.

83. Stern, 103.

84. *Vremya Novostei*, January 30, 2007, 1.

85. *Financial Times*, November 15, 2006, 3; December 19, 2006, 3; March 26, 2007, 16; *Oreanda RIA*, December 20, 2000; Prime-TASS, March 26, 2007 (20:48).

86. *Financial Times*, November 1, 2007, 3.

87. Gazprom Web site, March 15, 2007. Just to fill in the other pieces of this matrix, E.ON equity's ownership in Gazprom comes from a direct investment by E.ON when it bought up 3.5 percent of Gazprom stock and from an investment by Gerosgaz, a company that bought another 3 percent of Gazprom. To make it all the more confusing, Gerosgaz is a joint venture that E.ON owns with, believe it or not, Gazprom. In other words, Gazprom owns part of itself. That is why this part of what should be the main text is in a footnote.

88. Goldman, *The Piratization of Russia*, 163; Marshall I. Goldman, *Détente and Dollars* (New York: Basic Books, 1975), 134–36, 163.

89. *Ria Novosti*, March 20, 2007; *Eurasia Daily Monitor*, Jamestown Foundation (February 21, 2007). For a more complete list see Stern, 113.

90. *Financial Times*, February 4, 2006, 6; *Eurasia Daily Monitor*, Jamestown Foundation 4 (August 16, 2007).

91. ITAR TASS, June 22, 2006.

92. *Financial Times*, April 20, 2006, 4.

Chapter 7

1. *David Johnson's Russia List* 4, no. 127 (June 5, 2007).

2. *Financial Times*, May 29, 2007, 2.

3. *Argumenty i Fakty*, March 12, 2001, 3.

4. *Washington Post*, October 17, 2006, A1.

5. *New York Times*, October 18, 2006, A15; *Washington Post*, October 18, 2006, A8.

6. *Kommersant*, October 18, 2006.

7. *New York Times*, January 4, 2007, A1; *Kommersant*, October 18, 2006; *Washington Post*, December 31, 2005, A1.

8. *Washington Post*, October 17, 2006, 4.

9. Since these raids were conducted less than a month before the mid-term November national elections, Congressman Weldon charged it was a political attack designed to support his Democratic opponent. Given that the Republican Party controlled the Justice Department at the time, this seemed far-fetched. Yet it does add another layer of complexity to the charges that the Department of Justice under Attorney General Alberto Gonzales had at the time drawn up a list of local U.S. attorneys who were to be fired for failure to investigate corruption and election fraud by Democrats. The U.S. attorney in the Weldon case did, at

least, seem resolute given that the target was a ten-term Republican who, in the wake of all these scandals, went down to defeat.

10. *Financial Times*, April 9, 2007, 2; *New York Times*, March 5, 2007, A11.
11. *Wall Street Journal*, April 5, 2007, A2; *BP Statistical Review*, 30.
12. *Financial Times*, April 9, 2007, 2; *New York Times*, March 5, 2007, A11; *BP Statistical Review*, June 2007, 24, 27, 30.
13. *Financial Times*, April 9, 2007, 1.
14. *Eurasia Daily Monitor*, Jamestown Foundation 4, no. 30 (February 12, 2007), 4.
15. *Financial Times*, January 5, 2008, 1, 4.
16. *Moscow Times*, March 5, 2007, *International Herald Tribune*, December 27, 2007, 3.
17. *David Johnson's Russia List*, March 7, 2007, no. 55, item 35; *Financial Times*, March 12, 2007, 11.
18. *Financial Times*, November 29, 2007, Special Section, 3.
19. *Financial Times*, July 6, 2007, 6.
20. Leslie Dienes, "Natural Gas in the Context of Russia's Energy System," *Demokratizatsiia*, 15, no. 4 (Fall 2007), 408.
21. *Moscow Times*, March 5, 2007.
22. *Kommersant*, December 25, 2006, 8.
23. *Moscow Times*, March 5, 2007; Stern, 24.
24. *Financial Times*, June 1, 2007, 16; *Eurasia Daily Monitor*, Jamestown Foundation 4, no. 114 (June 19, 2007), 1.
25. *Moscow Times*, February 27, 2007.
26. *Moscow Times*, February 27, 2007.
27. *Financial Times*, August 31, 2007, 3.
28. Meeting of the Valdai Hills Discussion Group, Moscow, September 7, 2006; *Moscow Times*, February 27, 2006.
29. Grace, 194.
30. *Financial Times*, March 21, 2007, 21.
31. *New York Times*, July 13, 2006, A10.
32. *Moscow Times*, March 1, 2007.
33. *Kommersant*, December 25, 2006, 8.
34. *FC Novosti, Russian Financial Central Monitoring*, August 24, 2007.
35. *Kommersant Russia Daily Online*, August 31, 2007; Kremlin.ru, April 26, 2007.
36. *Moscow Times*, July 18, 2007.
37. Judy Dempsey, "In Russian energy plan, coal is a question mark," *International Herald Tribune*, December 27, 2007, 3.
38. Marshall I. Goldman, "The Russian Disease," *International Economy* (Summer 2005), 27.
39. *Moscow Times*, March 3, 2005; March 4, 2005.
40. *Moscow Times*, March 3, 2005.
41. *Moscow Times*, March 4, 2005.
42. *Moscow Times*, March 4, 2005.
43. Olga Kryshtanovskaia and Stephen White, "Putin's Militocracy," *Post-Soviet Affairs* 19, no. 4 (2003), 294.

44. *Kommersant*, November 30, 2007.

45. *New York Times*, December 12, 2007, C10.

46. Kremlin.ru, September 9, 2006, Transcript of meeting with participants in the third meeting of the Valdai Hills Discussion Club; Putin referred to Andrei Shleifer and Jonathan Hay.

47. V. Kleiner, "Korporativenoe upravlenie i effek"tivnost deliatel'nost kompani," *Voprosy ekonomikii*, no. 3 (2006), 98.

48. Michael D. Cohen, "Russia and the European Union: An Outlook for Collaboration and Competition in European Natural Gas Markets," *Demokratizatsiya*, 15, no. 4 (Fall 2007), 381.

49. *Vremya novosti*, August 31, 2007, 7; *Moscow Times*, October 4, 2007.

50. *Novaya Gazeta*, January 18, 2007, 2; *Noviye Izvestiai*, February 1, 2007, 5.

51. *Financial Times*, December 1, 2006, 4; *Rossiiskaia Gazeta*, October 3, 2005.

52. *Chicago Tribune*, September 27, 2006, 17.

53. *Financial Times*, December 12, 2007, 9.

54. *Wall Street Journal Asia*, November 9, 2007, 11.

55. *Izvestia*, November 27, 2007.

56. *Businessweek.com*, March 6, 2007; *David Johnson's Russia List* 55, no. 33 (March 7, 2007).

57. *Financial Times*, February 4, 2006, 6.

58. Other purchases include Severstall's acquisition of Rouge Industries in Dearborn, Michigan, and Lucchine Steel in Italy. Other recent purchases or green field construction include the Magnitigorsk Iron and Steel company's decision to build a cold rolled steel mill in Ohio; the EVRAZ purchase of Oregon Steel, Highveld Steel, and Vanadium; Norilsk Nickel's purchase of the Stillwater Mining Co. in Montana; the LionOre Canadian nickel mining company as well as the OM Mining Company; and Gazprom's purchase of the British gas marketing company, Pennine Natural Gas along with its affiliate, Natural Gas Shipping Services and what someday is expected to be the purchase of the much larger Centrica, which in turn owns British Gas. *Wall Street Journal*, July 17, 2007, A2; *Financial Times*, July 19, 2007, 1; *Moscow Times*, July 17, 2007; *New York Times*, September 6, 2007, C4.

59. *Financial Times*, September 21, 2007, 10; *Eurasia Daily Monitor*, Jamestown Foundation 4, no. 189 (October 12, 2007).

60. *Financial Times*, September 20, 1997, 4.

61. *Financial Times*, November 22, 2006, 2.

62. *China Daily*, September 17, 2007, 6; *Financial Times*, September 21, 2007, 3.

63. *Kremlin.ru*, February 10, 2007; *New York Times*, February 11, 2007, 4.

Glossary of People and Companies

People

Jack Abramoff A lobbyist in Washington who worked with Texas Congressman Tom DeLay and ended up in prison. One of their clients was a Russian energy company.

Roman Abramovich A former partner of Boris Berezovsky, Abramovich ended up as the main owner of Sibneft, which he sold to the state, making him the richest man in Russia. He used some of the funds to buy the Chelsea soccer team in London.

Vagit Alekperov A former minister of the petroleum industry in the Soviet Union who set aside valuable oil properties for himself and created LUKoil. He is the largest individual stockholder. LUKoil sold 20 percent of its stock to ConocoPhillips.

Svetlana Bakhmina A junior lawyer working for Yukos, she was arrested in the early morning hours and held hostage in an effort to force her boss to return to Russia for questioning after he fled to London.

Stanislav Belkovsky An analyst who works closely with Kremlin officials and who often has leaked information which signaled measures that were about to be taken by the Kremlin.

Boris Berezovsky One of the original oligarchs who became very close to members of Yeltsin's family. Among other assets he controlled were Sibneft and Aeroflot as well as ORT, the main state-owned TV network. Early on, he befriended Putin and helped him rise to power. However after being criticized on ORT, Putin turned on him, and Berezovsky fled to London where he lives in exile.

Leonard Blavatnik A Russian émigré with an MBA from the Harvard Business School. He is the principle owner of Access Industries, a U.S. company, which is a major stockholder in Tyumen Oil and SUAL, Russia's second largest aluminum company.

Sergei Bogdanchikov The CEO of Rosneft, the state-dominated oil company that took over ownership of the most valuable properties from Yukos.

Vladimir Bogdanov A veteran oil official who became CEO of Surgutneftegaz when it became privatized.

William Browder The grandson of Earl Browder, the head of the U.S. Communist Party. He established the Hermitage Capital Management Fund, which became a major investor in Gazprom and other Russian companies. After he criticized Russian corporate business practices, he was denied a Russian visa and prevented from returning to Russia.

Lord John Browne The CEO of British Petroleum (BP) who created a 50/50 partnership with Tyumen Oil.

Aleksandr Bulbov A lieutenant general in the Federal Narcotics Control Service who despite his rank was arrested by the FSB in what was thought to be a fight between government agencies over control of state assets.

Vladimir Butov The governor of the Nenets Autonomous District, an area rich in oil deposits, who has been charged with extortion and questionable practices by oil companies seeking to operate in the region.

William Casey The director of the CIA under President Ronald Reagan who is said to have worked with Saudi Arabia to increase oil production in an effort to precipitate a drop in oil prices and hurt the USSR's export earning capacity in order to bring about the collapse of the communist state.

Viktor Chernomyrdin The former minister of the gas industry who transformed the ministry into the joint stock company Gazprom. He later became a prime minister of Russia. Yeltsin later fired him and made him chairman of Gazprom.

Oleg Deripaska An oligarch who became a favorite of Putin, Deripaska won control of Rusal, the country's largest aluminum company. A controversial figure, at various times he has been denied visas to visit the United States. He is said to have become a major holder of General Motors stock.

Robert Dudley The managing head of the TNK-BP oil company.

Boris Fedorov A former minister of finance, he went on to become a major partner in United Financial Group. A major stockholder in Gazprom, Federov led an effort to remove its then CEO Rem Vyakhirev.

Dmytro Firtash A Ukrainian businessman who began by bartering goods between Ukraine and Turkmenistan, he became the head of Eural Trans Gas, a shadowy intermediary between Ukraine, Turkmenistan, and Gazprom.

Mikhail Fridman One of the original oligarchs who created Alfa Bank. He also became one of the principle owners of Tyumen Oil. He is one of the few original oligarchs who has survived the Putin purges.

Ivan Fursin A junior partner with Dmytro Firtash in RosUkrEnergo, the opaque intermediary that sold gas to the Ukrainian utility which supplies Ukrainian households.

Yegor Gaidar The acting prime minister during Yeltsin's first year as president of Russia. One of the architects of Russian shock therapy.

Viktor Gerashchenko The head of the Soviet and then the Russian Central Bank.

Vladimir Gusinsky One of the early oligarchs who created Most Bank and Media-Most, which became Russia's first private TV network. After his NTV network attacked Putin, he was arrested and eventually fled into exile.

Mikhael Gutseriev The founder of the oil company Russneft. He was forced to sell the company to Oleg Deripaska after the government issued a warrant for Gutseriev's arrest.

Ferenc Gyurcsany The prime minister of Hungary who is torn between joining with Russia or non-Russian groups in building a gas pipeline which would originate in the Caspian and Black Seas and transit through Europe.

Tony Hayward The successor to Lord John Browne as CEO of BP.

Mikhail Khodorkovsky Another of the original oligarchs who created the Menatep Bank, which in turn gained ownership of Yukos. Khodorkovsky was subsequently arrested and sentenced to 8 years in jail and Yukos was seized by the state.

Sergei Kiriyenko The prime minister of Russia from March 1998 until the financial crash of August 1998. Subsequently Putin appointed him the chairman of the Federal Atomic Energy Agency.

Helmut Kohl Chancellor of Germany from 1982 to 1998.

Alexander Korzhakov A KGB general who in 1985 became the head of presidential security when Boris Yeltsin was president. He was removed from office in 1996.

Konstantin Kosachëv Chairman of the International Affairs Committee of the Duma.

Alexei Kudrin Worked with Putin in the governor's office in St. Petersburg and later accompanied Putin to Moscow to work in the central government. A technocrat, he eventually became the minister of finance.

Platon Lebedev A partner of Khodorkovsky in Menatep and Yukos who was also found guilty of tax evasion and sentenced to jail.

Alexander Litvinenko A former agent of the KGB who fled to London and was subsequently poisoned.

Andrei Lugovoi A former KGB agent who was accused of poisoning Litvinenko and who refused to return to London after he was elected to the Duma as a member of the Liberal Democratic Party.

Alexander Lukashenko The president of Belarus who some have described as the last dictator of Europe.

Igor Makarov A bicycle champion from Turkmenistan who founded ITERA, which started out as a trading company and at one point became the second largest producer of natural gas in Russia. Its headquarters are in Jacksonville, Florida.

Enrico Mattei CEO of the Italian energy company Eni.

Valentina Matviyenko The governor of St. Petersburg.

Alexander Medvedev Deputy Chairman of Gazprom.

Dmitry Medvedev Chairman of Gazprom and for a time Director of the Presidential Administration in the Kremlin and subsequently first deputy prime minister, who also worked with Putin in St. Petersburg when he was deputy governor.

Alexei Miller The CEO of Gazprom who worked with Putin when he was deputy governor of St. Petersburg.

Bruce Misamore Chief financial officer of Yukos, an American who previously worked for Marathon Oil in the United States.

Semion Mogilevich A shadowy figure accused by the FBI of criminal activity and being a mafia leader who is thought to be involved in the sale of gas to Ukraine.

Nursultan Nazarbayev The president of Kazakhstan.

Leonid Nevzlin A close friend and collaborator of Khodorkovsky who fled in exile to Israel before Khodorkovsky was arrested and Yukos was seized by the state.

Saparmurat Niyazov The leader of Turkmenistan until his death.

Ludwig and Robert Nobel Brothers who were among the first to develop Russia's oil fields around Baku before World War I and the Revolution.

Nikolai Patrushev Head of the FSB, the successor to the KGB.

Vladimir Petukhov The mayor of an oil-rich city in Siberia where many Yukos operations were located. After complaining about Yukos's failure to pay its taxes, he was found murdered.

Evgeny Primakov A former head of the KGB who was appointed prime minister in September 1998 after the financial collapse and who was removed in May 1999.

Lee Raymond CEO of Exxon-Mobil.

John D. Rockefeller One of the early developers of the oil industry in the United States and the founder of Standard Oil.

Leonid Roketsky The governor of the Tyumen region who at the same time was chairman of the Tyumen Oil Company.

Rothschild brothers International bankers and early investors and developers of oil production in the Baku region.

Mikhail Saakashvili The president of Georgia who earlier attended Columbia University in New York.

Gerhard Schroeder The chancellor of Germany who promoted the building of Nord Stream, a Russian-German pipeline, and then became the chairman of the board of directors.

Igor Sechin A former KGB agent who became deputy chairman of the Kremlin administration while simultaneously serving as chairman of the Board of Directors of Rosneft.

Igor Shuvalov An economic adviser to Putin.

Oleg Shvartsman A shadowy figure who runs the $36-billion Finansgroup Investment Fund. This fund is reputed to manage the assets of high-ranking government officials who have funneled government assets into their own accounts.

Alexander Smolensky An early oligarch who created the SBS/AGRO bank and with Berezovsky became an owner of Sibneft.

Anatoly Sobchak Putin's professor in law school who became governor of St. Petersburg and appointed Putin as his deputy.

Sergei Stepashin The former head of the FSB who served as prime minister from May 1999 to August 1999 and subsequently became head of the Duma Audit Chamber.

Sergei Storchak The deputy minister of public finance in charge of administering and investing the country's stabilization fund. He was arrested on charges of embezzlement in late 2007 as part of what was thought to be an effort by some of the siloviki to gain control of the billions of dollars held in that fund.

Steven Theede An American who worked for ConocoPhillips and was later appointed as chief operating officer of Yukos.

Gennady Timchenko A long-time friend of Putin who with Putin is rumored to share the ownership of Gunvor, a company selling petroleum.

Andrei Vavilov A former deputy finance minister who became the predominant owner of Northern Oil, a company he eventually sold to Rosneft for a very expensive price.

Viktor Vekselberg An early oligarch who became a major partner in Renova, which has major holdings in Tyumen Oil. He also became one of the owners of SUAL, one of the country's aluminum manufacturers. He also financed the purchase of the Fabergé eggs so they could be returned to Russia and paid for the return of the bells of the Danilov Monastery from Harvard University.

Rem Vyakhirev The president of Gazprom until he was not reappointed in 2001. Formerly he was the Deputy Minister of the Gas Industry.

Matthias Warnig A former Stasi Secret Police agent who befriended Putin when they were both stationed in East Germany. Waring later ran the Dresdner Bank office in St. Petersburg and was selected to head the Nord Stream pipeline project in the Baltic Sea.

Kurt Weldon A ten-term congressman from Pennsylvania whose daughter became the public relations principal for ITERA.

Grigory Yavlinsky An economist who worked for both Yeltsin and Gorbachev and who later became the head of the Yabloko political party.

Viktor Yushchenko The president of Ukraine.

Gennady Zyuganov The head of the Communist Party in Russia.

Companies

Arcelor A Benelux steel company that for a time considered forming a partnership with a Russian company.

Blue Stream A natural gas pipeline from Russia to Turkey under the Black Sea.

Eni / Ente Nazionale Idrocarburi An Italian energy company that became a major purchaser of Soviet oil and gas.

E.ON A German natural gas company that bought up Ruhrgas and has partnered with Gazprom in several projects.

Gasunie A Dutch pipeline company that is partnering in the construction of the Nord Stream gas pipeline in the Baltic Sea.

GECF The Organization for Gas Exporting Countries which is a forum for gas producers. It as yet lacks the powers of an OPEC-type organization.

ITERA For a time, Russia's second-largest producer of natural gas. It is headquartered in Jacksonville, Florida.

Kharyaga An oil field located in Timan-Pechora, a northern province. The French company Total has the operating concession.

Kovykta A gas field located in northern Siberia. BP has the operating concession, but it failed to fulfill the terms of the production agreement with the state. As a penalty it was forced to provide an ownership share to Gazprom.

LUKoil A private oil company put together by Vagit Alekperov. ConocoPhillips now owns 20 percent of its stock. LUKoil purchased the Getty Oil filling station network.

Menatep The bank created by Khodorkovsky.

MOL A Hungarian natural gas utility.

NABUCCO A pipeline that the European Union is seeking to build as a way of bypassing Russian-controlled gas pipelines to Europe.

Nefteyugansk One of the main producing sites for Yuganstneftegas which was the main producing unit for Yukos until it was taken over by Rosneft.

NEGP The North European Gas Pipeline, now called Nord Stream.

Nord Stream The pipeline Gazprom is building in the Baltic Sea connecting Russia to Germany designed to bypass Poland and Ukraine.

Norex A Canadian oil development company that was pushed out of its development work in Russia by Tyumen Oil.

Norilsk Nickel A major producer of nonferrous metals controlled by Vladimir Potanin.

OGEC The Organization of Gas Exporting Countries, a possible OPEC.

OMV An Austrian utility company which is seeking control of MOL.

OneksimBank The bank formed by Vladimir Potanin.

Renova A U.S. company controlled by Blavatnik and Vekselberg which has major holding in TNK and SUAL, among others.

Romaskino For a time one of the world's largest oil wells.

Rosneft The state-owned oil company that took over most of Yukos's assets.

Ruhrgas A German natural gas distribution company that early on cooperated with Gazprom and owns shares in Gazprom. It was bought up by E.ON.

Samotlor A major oil-producing site in Russia.

SEGP A natural gas pipeline Gazprom is seeking to build from Turkey to Western Europe.

Sibneft The Russian oil company privatized by Berezovsky, who in turn brought in Abramovich, who in turn sold it to Gazprom.

South Stream Yet another natural gas pipeline which if built would transport gas in Southern Europe and compete with NABUCCO.

TNK-BP A 50/50 joint venture formed by BP and Tyumen Oil.

Tyumen Oil (TNK) One of the privatized oil companies; controlled by Fridman, Blavatnik, and Vekselberg, which formed a joint venture with BP.

Volga-Urals One of the oil fields developed along the Volga.

Wingas A natural gas joint venture formed between Wintershall, the German company, and Gazprom.

Wintershall A German natural gas distribution which has entered into several joint ventures with Gazprom.

Yukos For a time, the largest privatized oil company in Russia until it was taken over from Khodorkovsky; much of it was taken over by Rosneft.

Index

Rising Powers, Shrinking Planet
How Scarce Energy is Creating a New World Order
Michael Klare

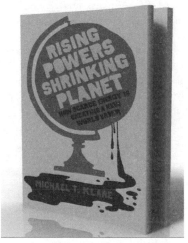

As essential sources of energy rapidly deplete, governments rather than corporations are taking over the pursuit of resources. In a radically altered world - where a resurgent Russia holds Europe to ransom, and the West is forced to compete with the 'Chindia' juggernaut - Michael Klare, the pre-eminent expert on resource geopolitics, forecasts a future of surprising new alliances and explosive danger.

9781851686483 - HB - £18.99
9781851686285 - PB - £10.99

"A brilliant exposition on one of the gigantic problems facing society. Klare is a top expert on the politics of energy and resources. Read him!"—**Paul R. Ehrlich,** author of *The Dominant Animal*

Michael Klare is the author of thirteen books, including *Blood and Oil* and *Resource Wars*. A regular contributor to *Harper's, Foreign Affairs*, and the *Los Angeles Times*, he is the defense analyst for *The Nation* and the director of the Five College program in Peace and World Security Studies at Hampshire College in Amherst.

Browse further titles at **www.oneworld-publications.com**

Oil: A Beginner's Guide
Vaclav Smil

"In a fluent, easy style [Smil] delves into the world of oil from its discovery on the ground through to its effect on prices at the petrol pumps, and to its impact on future generations." - *The Good Book Guide*

"Smil's knowledge is famously and fabulously encyclopedic." - **Professor Jeffrey D. Sachs,** Director, Earth Institute

9781851684526 - PB - £9.99

Energy: A Beginner's Guide
Vaclav Smil

A concise, accessible guide both to energy and to global warming and our efforts to prevent it.

"Smil's 'Energy' is rich in thoughtful insights and written in sparkling prose; this little book offers a sweeping survey of the many ways that energy moves our economy and ecology." - **David G. Victor,** Director, Program on Energy and Sustainable Development, Stanford University.

"Admirably clear and comprehensive" - **Sir Crispin Tickell,** Former Chairman of the International Institute for Environment and Development.

9781851684522 - PB - £9.99

Browse these and other titles at www.oneworld-publications.com